A PHILOSOPHY FOR THE SCIENCE OF ANIMAL CONSCIOUSNESS

This book attempts to advance Donald Griffin's vision of the "final, crowning chapter of the Darwinian revolution" by developing a philosophy for the science of animal consciousness. It advocates a Darwinian bottom-up approach that treats consciousness as a complex, evolved, and multidimensional phenomenon in nature rather than a mysterious all-or-nothing property immune to the tools of science and restricted to a single species.

The so-called emergence of a science of consciousness in the 1990s has at best been a science of human consciousness. This book aims to advance a true Darwinian science of consciousness in which its evolutionary origin, function, and phylogenetic diversity are moved from the field's periphery to its very centre, thus enabling us to integrate consciousness into an evolutionary view of life. Accordingly, this book has two objectives: (i) to argue for the need and possibility of an evolutionary bottom-up approach that addresses the problem of consciousness in terms of the evolutionary origins of a new ecological lifestyle that made consciousness worth having and (ii) to articulate a thesis and beginnings of a theory of the place of consciousness as a complex evolved phenomenon in nature that can help us to answer the question of what it is like to be a bat, an octopus, or a crow.

A Philosophy for the Science of Animal Consciousness will appeal to researchers and advanced students interested in advancing our understanding of animal minds as well as anyone with a keen interest in how we can develop a science of animal consciousness.

Walter Veit is a postdoctoral research associate at the University of Bristol. His interests stretch widely across science and philosophy, but they are primarily located at the intersection of the biological, social, and mind sciences in addition to empirically informed philosophy and ethics.

In *A Philosophy for the Science of Animal Consciousness*, Walter Veit argues for a more thoroughly Darwinian approach to understanding how consciousness has come into existence, and in which beings it is to be found. By shifting the focus away from human consciousness, he helps us to understand the diversity of forms of consciousness that exist in other animals.

Peter Singer, *Ira W. DeCamp Professor of Bioethics, Princeton University*

The a priori position that human consciousness differs from that in other animals has become hard to defend. In an eye-opening account, Walter Veit explains why. As a true philosopher, he delves into both the richness and ambiguity of the concept of consciousness.

Frans de Waal, *C. H. Candler Professor of Psychology, Emory University*

Walter Veit takes a deep historically- and empirically-informed look at the origins of cognitive ethology and re-centers the field on Donald Griffin's original idea that it's consciousness in animals that really matters. He tackles the question of how to fit an account of consciousness into the life histories of individual animals, using a Darwinian framework that emphasizes the variety and adaptive radiation of forms and functions of consciousness in the evolutionary tree.

Colin Allen, *Distinguished Professor, University of Pittsburgh*

Some think that explaining consciousness is beyond the scope of evolutionary theory. Undeterred, Walter Veit rolls up his sleeves and gets on with the task. Drawing on the latest work in evolutionary biology, cognitive ethology, and neuroeconomics, he reverse engineers consciousness, distinguishing its different dimensions and components and identifying its roots in an ancient evaluative system which evolved to manage the complex action-selection problems faced by early forms of animal life. This is a pioneering and important book, which is informed throughout by an awareness of the rich diversity of animal life and experience. It will challenge your view of consciousness and transform your attitude to your fellow creatures.

Keith Frankish, *Honorary Professor, Department of Philosophy, University of Sheffield*

Walter Veit has written a very thoughtful and thought-provoking philosophical exploration of the evolutionary origins of consciousness. He aims to bring us closer to a true biological science of animal consciousness, what Donald R. Griffin, the founder of the field of Cognitive Ethology, termed the "crowning chapter of the Darwinian revolution." That chapter is to be written by studying the mental experiences of animals in their daily lives and natural worlds.

Veit's work exhibits the fruitfulness of the growing collaborations between philosophers and scientists of animal behavior to the clear benefit of both. He gives serious consideration to the problem of consciousness and the evolution of forms of consciousness, integrating the work of many disciplines and delineating the likely functional significance of consciousness and its varieties in different species.

Veit offers persuasive arguments and examples that evaluative consciousness lies at the core of the phenomenon of consciousness, though leaving enough to argue about and discuss fruitfully as to other characteristics that may be strong contenders for that role.

His work is a significant contribution to the field and well worth delving into.

Carolyn A. Ristau

A PHILOSOPHY FOR THE SCIENCE OF ANIMAL CONSCIOUSNESS

Walter Veit

Routledge
Taylor & Francis Group
NEW YORK AND LONDON

Designed cover image: Photograph by Francesca M. Cornero

First published 2023
by Routledge
605 Third Avenue, New York, NY 10158

and by Routledge
4 Park Square, Milton Park, Abingdon, Oxon, OX14 4RN

Routledge is an imprint of the Taylor & Francis Group, an informa business

Library of Congress Cataloging-in-Publication Data
Names: Veit, Walter (Researcher in philosophy), author.
Title: A philosophy for the science of animal consciousness / Walter Veit.
Description: First edition. | New York, NY : Routledge, 2023. | Includes bibliographical references and index.
Identifiers: LCCN 2022061465 (print) | LCCN 2022061466 (ebook) | ISBN 9781032343600 (hbk) | ISBN 9781032343617 (pbk) | ISBN 9781003321729 (ebk)
Subjects: LCSH: Consciousness in animals.
Classification: LCC QL785.25 .V45 2023 (print) | LCC QL785.25 (ebook) | DDC 591.5--dc23/eng/20230320
LC record available at https://lccn.loc.gov/2022061465
LC ebook record available at https://lccn.loc.gov/2022061466

ISBN: 978-1-032-34360-0 (hbk)
ISBN: 978-1-032-34361-7 (pbk)
ISBN: 978-1-003-32172-9 (ebk)

DOI: 10.4324/9781003321729

Typeset in Bembo
by KnowledgeWorks Global Ltd.

I dedicate this book to my parents Ieva and Werner.

CONTENTS

FIGURES AND TABLE

Figures

Table

PREFACE

In Search of the Place of Mind in Nature

The target phenomenon of this monograph is one that could hardly constitute a greater challenge to a paleobiologist. It is a phenomenon that is said to leave no fossil trace and has repeatedly been described as the hardest problem of biology: consciousness. In nature, we seem to find a striking difference between systems without any sort of conscious experience, like Australian bush fires, the plants that succumb to them, robots, bacteria, our planet, other stars, and the universe as a whole, and those systems, like humans, for which there is something it is like to be them – or so most take for granted. When we ask whether there is something it is like to be a bat, or any other living organism for that matter, we are asking both whether they have subjective experiences (of any kind) and what these experiences consist in. Yet, how could we possibly learn about the nature of this elusive phenomenon?

The problems of consciousness have for a long time puzzled both scientists and philosophers, even deemed exceedingly difficult if not impossible to answer: What is consciousness and why does it exist at all? Could consciousness come in degrees and different variations or is it like a light switch that is either "on" or "off"? Finally, which animals are conscious and do they differ in their subjective experiences? Are humans the only conscious beings on our planet? Or should we include all mammals? Birds as well? Or all the animals? Why not say that all life is sentient? This view is sometimes called *biopsychism* and though it will strike many as surely too strong, that does not mean that it lacks defenders.[1]

The German biologist, philosopher, and artist Ernst Haeckel (1892) – who is sometimes described as the "German Darwin" for his devout defence and development of Darwin's theoretical framework in Germany[2] – was the one who coined the term "biopsychism". But he eventually went on to defend an even broader view called *panpsychism*: the view that "[a]ll matter is ensouled" and that feeling should be conceived of as "a universal world-principle" (1892, p. 483). Contemporary panpsychists think that our fundamental scientific image of physics needs to be radically updated to include an aspect or degree of mentality in all matter, such as electrons, in order to make sense of the presence of minds, a view that even its proponents admit is readily rejected by most philosophers and non-philosophers alike as – to put it bluntly – absurd (Goff et al., 2020). Yet, the view has very prominent defenders such as Thomas Nagel and David Chalmers, who have been incredibly influential in the shaping of the philosophical and scientific discourse around consciousness and urge us to take this radical option seriously. They hold that the problems of panpsychism are no more serious than those for any other view, though Nagel (1986) also admits that the view has "the faintly sickening odor of something put together in the metaphysical laboratory" (p. 49). But while it appears easy enough to dismiss such radical views as going too far, it is also apparent that there is little agreement on how we could even *possibly* settle the question.

A Wild Sweep of Alternatives

Disagreement about the possibility of pain in fish or invertebrates such as insects sometimes appears no less contested than the metaphysical view that consciousness pervades the universe. Largely, this is due to a worry the British biologist Thomas Huxley – also known as *Darwin's bulldog* – once famously expressed:

> [W]hat consciousness is, we know not; and how it is that anything so remarkable as a state of consciousness comes about as the result of irritating nervous tissue, is just as unaccountable as the appearance of the Djin when Aladdin rubbed his lamp in the story, or as any other ultimate fact of nature.
>
> *Thomas Henry Huxley in Huxley and Youmans (1868, p. 178)*

Huxley did not perceive how we could possibly explain consciousness as a causally efficacious materialist phenomenon, which led him to endorse epiphenomenalism, i.e. the dualist view that subjective experience is only the effect, never the cause of the physical processes of the brain.[3] His concerns were later better articulated by Joseph Levine (1983), who expressed the mind-body problem in terms of what he called "the explanatory gap" between the mental and the physical – an epistemological rather than ontological gap he thought could only be bridged by eliminating the mental.[4] That would be a radical non-dualist answer to the mind-body problem that many would consider to be too "hard-headed", reductive materialism gone too far. How could we possibly eliminate the most cherished and directly experienced aspect of our mental lives? The framing now more commonly used in debates about consciousness is Chalmers's (1995) description of this explanatory gap as the so-called hard problem of consciousness. Chalmers maintained that while we can readily make progress on the "easy problems of consciousness", i.e. the functional, computational, and mechanistic side of the mind, using the standard tools and methods of cognitive science, none of this appears to address the hard problem of how it generates a first-person phenomenological feel of mental phenomena:

> What makes the hard problem hard and almost unique is that it goes *beyond* problems about the performance of functions. To see this, note that even when we have explained the performance of all the cognitive and behavioral functions in the vicinity of experience – perceptual discrimination, categorization, internal access, verbal report – there may still remain a further unanswered question: *Why is the performance of these functions accompanied by experience?* A simple explanation of the functions leaves this question open.
>
> *David Chalmers (1995, p. 202) [emphasis in original]*

This framing of the problem of consciousness makes it appear as if science could never address the qualitative feel, i.e. the "qualia", of subjective experience. Following Nagel's (1974) famous essay "What Is It Like to Be a Bat?", in which he set out to argue that while bats are likely conscious we could never know what their subjective experience was like, phenomenological properties or qualia are now typically treated as something like a second-order property of what it is like or what it feels like to have a mental state, as opposed to a first-order property of the state itself (Sytsma & Machery, 2010, p. 299). Thanks to Chalmers' and Nagel's influence on the philosophy of mind, it is now typical to consider the problem of consciousness as identical to the problem of qualia rather than of a particularly rich cognitive phenomenon with qualitative aspects that may be unique to humans, as it was in the literature of the philosophy of mind in the 1980s (Godfrey-Smith, 2016a, 2016c, 2017a). Could this combination of two formerly distinct problems have given rise to the increasing conviction among many philosophers and scientists that there is something like a hard problem that cannot be solved?

An even more radical rejection of the idea that science could provide a materialist account of consciousness can be found in the form of *idealism*, i.e. the romanticist notion that everything is mental. What are we to make of this, what Godfrey-Smith (2020b) candidly called a "wild sweep of these alternative views of the universe" (p. 14)? In thinking about the place of mind in nature, there almost appears to be a dreadful possibility that *anything goes* and this is not restricted to "mere" philosophical discussions about the very nature of mind. If we ask about the presence of other minds in non-humans, it appears that we are faced with just as much uncertainty as with the big-picture view about the relationship between matter and mind, at least two senses in which we can ask for the place of mind in nature.

How are we to respond to a biopsychist who points to the autonomy, sophisticated sensory feedback, and decision-making of single-celled organisms? To assert that all of the actions of an animal could be explained by mere mechanics without the presence of mind is a tactic frequently used in discussions over all kinds of possible boundaries, including fish, insects, and mammals. Why is someone wrong who denies pain in octopuses or crabs? How could we possibly settle these debates about where to draw the line? Many answers to this question of the place of mind in nature have been proposed, such as the eliminativist or illusionist view that *no one* has consciousness in the sense of possessing qualia (Dennett, 1991; Frankish, 2017; Levine, 1983), the exclusive attribution of consciousness to humans (Macphail, 1998), only to the great apes (Bermond, 2001), only to mammals and birds (Edelman & Tononi, 2000), to all mammals, birds, and non-avian reptiles (Cabanac et al., 2009), to all vertebrates (Mashour & Alkire, 2013), to all vertebrates as well as some invertebrate groups such as cephalopods, crustaceans, and insects (Barron & Klein, 2016; Bronfman et al., 2016; Ginsburg & Jablonka, 2010; Tye, 2016;), to plants as well (Gagliano, 2017; Trewavas et al., 2020), to all living organisms including single-celled ones (Margulis, 2001; Reber, 2019), and to all entities in the universe (Goff et al., 2020).[5] Views on the presence of consciousness range from none to all.

We are faced with such a diversity of alternative models of consciousness that it almost seems like we have what has in similar contexts of abundance been called an "embarrassment of riches" (Veit, 2021d). Without a *standard* for thinking about these problems of the mind, it almost looks like, as Godfrey-Smith (2020b) put it, "[p]eople can say whatever they like" (p. 15). This view of philosophy as a state of indefinite arbitrariness is to be strongly resisted. Following a long tradition of naturalist thinkers, this book firmly rejects the view common in some areas of philosophy that our profession is primarily engaged in a game of mere conceptual exploration; that philosophy is merely concerned with expanding the space of *possible* views.[6] The following chapters constitute an exercise in naturalistic philosophy to make sense of the place of consciousness in nature by providing the science of consciousness with a much needed *standard* that is unfortunately still lacking.

A Darwinian Standard

As other Darwinian thinkers have argued, this standard should not be the cherished insight derived from human first-person experience, but the modern twenty-first century theory of evolutionary biology, the one theory that a biological approach to consciousness should not be neutral towards. It is only by investigating the evolutionary origins of consciousness and the ecological lifestyles of these first conscious entities that we will truly understand the place of consciousness in nature without being misled by the particularities, idiosyncrasies, and complexities of the human mind. The shared ancestry of all life on Earth provides us with a rich set of theoretical tools and constraints with which to understand the origins of biological phenomena. And of course, consciousness is just that, an evolved biological phenomenon – something that is now widely accepted among both philosophers and scientists writing about consciousness.

So one would be led to believe, as the neuroscientist Simona Ginsburg and evolutionary biologist Eva Jablonka note in their recent 2019 book *The Evolution of the Sensitive Soul*, that philosophers and scientists alike had firmly integrated evolutionary theory into "the framework of consciousness studies, both as a yardstick for measuring the validity of new theories and as a source of insights" (p. x). Indeed, Darwin himself had already realized 21 years prior to the publication of *The Origin of Species*, in a little private notebook,[7] a view of life in terms of shared ancestry would radically transform our view of nature and our place within it. As he put it: "Origins of Man now proved. – Metaphysics must flourish. – He who understands baboon would do more towards metaphysics than Locke" (Darwin, 1838, p. 84e). But, despite the efforts of many Darwinian thinkers, there has been no biological revolution in our thinking about consciousness. Evolutionary and ecological thinking about the role of consciousness in nature has so far only played a surprisingly small role in the study of consciousness. Ginsburg and Jablonka describe this lack of Darwinian thinking within a supposedly naturalistic study of consciousness highly critically: "until very recently there has been a strange lacuna in the field. Although most scientists and philosophers who write about consciousness are now convinced that it is a biological process that is a product of evolution, its evolutionary origins are rarely central to their discussions" (2019, p. x).

Undoubtedly, this can be explained through the perceived difficulty and speculative nature of adaptationist reverse-engineering approaches to the mind – especially the human mind – which have been sneered at as "just-so stories": plausibly sounding explanations that aren't empirically testable (Gould & Lewontin, 1979). While there have been innumerable attempts at a functionalist approach to consciousness, due to its status as a standard objection to epiphenomenalist views that make its evolution a mystery,[8] such thinking has unfortunately often avoided an investigation of its evolutionary origins in other animals, instead focusing on humans and humans alone.

Yet, it is no surprise that such evolutionary explanations would be avoided in a scientific investigation that was already seen as deeply suspect due to the lingering after-effects of the behaviourist project that banished consciousness from science. But this neglect of Darwinian thinking is unfortunate because evolutionary theory provides us with both a rich theoretical framework for thinking about consciousness and an important set of constraints that any theory of consciousness should account for. If we can build a theory of consciousness that doesn't leave its evolutionary origins a mystery, one that can explain the dawn of qualia, then we will no longer be in a position where people could say that no view is better than any other and all cards should be left on the table as equal contenders. A theory that can explain the evolution of consciousness in a gradual fashion through small incremental steps is to be preferred over any theory that demands a dualist carving of nature at its joints, a big jump or a sudden explosion of mindedness. Instead, we would end up with a historical explanation of the place of consciousness as a complex phenomenon in nature that at least substantially narrows the explanatory gap between matter and mind. As Ginsburg and Jablonka put it:

> Evolutionary theory is [...] the most general framework for understanding the biological world. It is a conceptual bottleneck through which any theory of life and mind must pass.

> If a biological (or psychological, or sociological) theory fails to pass through this bottleneck, it is likely there is something seriously wrong with it.
>
> *Simona Ginsburg and Eva Jablonka (2019, pp. ix–x)*

Like most scientists studying consciousness, this book will treat consciousness as a complex evolved biological phenomenon related to the brain and nervous system of animals; something that was *built* over aeons of evolutionary time. But if consciousness is a biological phenomenon, then it ought to be treated as such. Whereas some prominent figures such as Nagel (2012) see the problems of consciousness as a fundamental flaw in evolutionary theory, their views have shown a striking lack of knowledge and underutilization of the theoretical toolkit modern evolutionary biology has to offer.

If we are interested in the place of mind in nature, we must place the question of its origin, function, and phylogenetic diversity across the tree of life at the very heart of a true biological science of consciousness. It is only by asking the functionalist question of what consciousness in all of its varieties and gradations does *for* healthy sentient agents within their normal ecological lifestyles and the natural environments they have evolved in that we can transition towards a true biological study of consciousness. This naturalist endeavour is fundamentally what *A Philosophy for the Science of Animal Consciousness* will attempt to accomplish.

While there have been numerous attempts to address the problems of consciousness through a functionalist/evolutionary approach (many of which I will draw on throughout), my approach will stand out by offering a new strategy for making progress on these problems through an emphasis of animal life histories in addition to focusing on the healthy and pathological varieties and gradations of consciousness as a complex phenomenon in nature. Naturally, my specific proposals and theoretical sketches may turn out to be wrong, but it is only in attempting to integrate evolutionary and ecological thinking with the science of consciousness that we can truly move towards a study of consciousness as a widespread natural rather than merely human phenomenon. And to provide such a possibility proof for a bottom-up approach and evolutionary framework that can help us to think about the *raison d'être* consciousness has for organisms within their natural lives is the goal of this book.

Notes

1 Consider, for instance, Herbert S. Jennings (1904), Henri Bergson (1920), Maxine Sheets-Johnstone (1999), Lynn Margulis (2001), and Arthur Reber (2019).
2 See Aveling (1886) and Kutschera et al. (2019).
3 See Huxley (2011) and Campbell (2001) for a deeper discussion of Huxley's views.
4 Godfrey-Smith (2020b) contests whether Huxley expressed an early version of the explanatory gap. Yet, the parallels are quite striking. Whereas Huxley found himself led towards epiphenomenalism, Levine suggested that the only way to close the gap was to become an eliminative materialist in regard to the qualitative side of the mind, though admitting that it has so far remained stubbornly "resistant to philosophical attempts" at elimination (Levine, 1983, p. 361). They, nevertheless, both saw the mental side as a deeply problematic explanatory challenge without any intuitively attractive solutions.
5 This list is not meant to be exhaustive. See Liljenström and Århem (2008) for a collection of essays and Griffin (2001), Ginsburg and Jablonka (2010, 2019), and Dawkins (2021) for other expositions of alternative views on the phylogenetic distribution of consciousness that my list draws on.
6 I thank Kim Sterelny for raising this point against the panpsychist literature at a bonfire at the *Philosophy of Biology at Dolphin Beach* (PBDB) workshop in 2020.
7 See Notebook M (Darwin, 1838).
8 See the target article of Dawkins (1990) and the commentaries to it for an excellent discussion of epiphenomenalism and the causal role of subjective feelings.

ORIGINS OF THE CHAPTERS

Parts of this book contain previously published material. Chapter 1 is an extended version of "Veit, W. (2023b). Health, Consciousness, and the Evolution of Subjects. *Synthese*". Chapter 3 is a longer and more extensive version of "Veit, W. (2022c). The Origins of Consciousness or the War of the Five Dimensions. *Biological Theory*". Section 2.6 also reuses some material from this article. Chapter 4 is an expanded and improved version of "Veit, W. (2022a). Complexity and the Evolution of Consciousness. *Biological Theory*". Sections 1.1., 1.3, 6.3, and 6.4 also reuse some of the material from this article. Chapter 5 is an improved and expanded version of "Veit, W. (2022e). Towards a Comparative Study of Animal Consciousness. *Biological Theory*", with added discussions of octopuses, fish, reptiles, and birds. Section 1.1 also uses some of the content from this article. Finally, Section 3.2.2 also includes material from "Veit, W. (2022b). Consciousness, Complexity, and Evolution. *Behavioral and Brain Sciences, 45*, e61" and Section 5.4 uses some of the material from "Veit, W. (2021b). The Evolution of Knowledge during the Cambrian Explosion. *Behavioral and Brain Sciences, 44*, e47", both reproduced with permission from Cambridge University Press.

ACKNOWLEDGEMENTS

This book is the result of several years of thinking about consciousness and its evolution in Sydney, London, Munich, Cambridge, Oxford, and finally Bristol. While philosophy is often described as an ivory tower activity, this could be no further from the truth for naturalistic philosophy, and I have benefited greatly from the advice and feedback from scientists and philosophers alike. During this time, I have been in contact with more people than I could mention here, so I pre-emptively ask forgiveness from anyone I have forgotten to mention.

Since this book originated from my dissertation, I would like to thank all of my doctoral advisors for their help and consistently constructive criticisms. Firstly, my primary supervisor Paul Griffiths, who helped the development of my dissertation immensely with his wide range of expertise, humour, and ultimately funding from his ambitious "A Philosophy of Medicine for the 21st century" project that supported my PhD (for which I acknowledge funding from the Australian Research Council's Discovery Projects funding scheme [project number FL170100160]). Secondly, Peter Godfrey-Smith, whose work inspired me to write a dissertation on animal consciousness and whose feedback and extensive knowledge of the field have been incredibly helpful. Thirdly, I would like to thank Marian Dawkins, Professor of Animal Behaviour at the University of Oxford, who kindly agreed to serve as an external advisor for roughly the last half of the duration of my PhD. While we eventually gave up on formalizing this status due to excessive paperwork, she nevertheless deserves mention here for her role as an informal advisor and her helpful feedback on my dissertation.

Turning a dissertation into a book is far from an easy endeavour, and I have benefited greatly from exchanges with members of the University of Sydney. Being part of Griffiths' wonderful interdisciplinary Theory and Method in Biosciences group at the Charles Perkins Centre – one of Australia's leading medical research institutes – enabled me to present my work multiple times to an entire group of philosophers of biology and to grow into an interdisciplinary researcher with a firm understanding of evolution. Indeed, this book very much reflects the mission of this "philosophy lab" to employ the integrative power of biological theory and especially evolutionary biology to remove "conceptual and methodological roadblocks to the advancement of science".

Next, I would like to acknowledge four host institutions that I visited during my PhD. Firstly, I would like to thank Jonathan Birch, who hosted me for a term in his Animal Sentience lab within the Centre for Philosophy of Natural and Social Science at the London School of

Economics, which provided the perfect place to discuss the ideas of my dissertation. Secondly, I would like to thank Stephan Hartmann, who hosted me for roughly two months at the Munich Center for Mathematical Philosophy at the Ludwig-Maximilians-Universität München, where I delved deep into the possibility of a mathematical framework for consciousness, taught a course in the history and philosophy of the life sciences, and was eventually honoured with the status of external member. Thirdly, I would like to thank Nicola Clayton for letting me visit her Comparative Cognition Lab at the University of Cambridge, Department of Psychology, for almost half a year. Observing actual research on animal cognition and engaging with researchers at her lab have been incredibly helpful in writing my dissertation and this book. The featured bird on the cover is a common raven (*Corvus corax*) and belongs to the family of Corvidae, which will be discussed later in this book. I would very much like to thank Francesca Cornero – a PhD student at Clayton's Comparative Cognition Lab – for allowing me to use her photograph for the cover of this book. Lastly, I would like to thank Rob Salguero-Gómez for hosting me in his life-history team at the University of Oxford, Department of Zoology, and later the new Department of Biology, which helped me to dive into the depths of life-history theory.

I have also presented ideas of this book to audiences at meetings of the Association for the Scientific Study of Consciousness, the International Network for Economic Method, the International Society for the History, Philosophy, and Social Studies of Biology, the Australasian Association of Philosophy, the Joint Brazilian Annual Ethological/Latin American Ethological Conference, and the Power of Scents workshop Sokoine organized by APOPO at the University of Agriculture in Tanzania, which was followed by a Safari in the Mikumi National Park – the perfect place to think about the evolutionary function of consciousness. Furthermore, I also would like to thank audiences at a keynote I gave at the 16th Universities Federation for Animal Welfare (UFAW) Student Conference and research seminars at the Ludwig Maximilian University of Munich and the University of Bayreuth.

For comments, I also owe gratitude to Daniel Dennett, Nicola Clayton, Jonathan Birch, Kristin Andrews, David Spurrett, Don Ross, Colin Allen, Stefan Gawronski, Kate Lynch, Peter Takacs, Carolyn Ristau, as well as reviewers for the previously published article versions of my chapters. Furthermore, I would also like to thank Samir Okasha who has supported me both as a master's student at the University of Bristol in their Philosophy of Biological and Cognitive Sciences program and now as a postdoctoral research associate on his Representing Evolution project (for which I also acknowledge funding from the European Research Council [ERC] under the European Union's Horizon 2020 research and innovation programme [grant agreement number 101018533]). Returning to Bristol has provided me with the perfect place to turn my doctoral dissertation into an accessible, expanded, and improved monograph.

I would also like to express my thanks to anonymous reviewers at both Routledge and Palgrave Macmillan for their feedback on the drafts of this book. My editor Andrew Weckenmann likewise deserves my gratitude for his help and encouragement in turning my thesis into a book accessible to a wide audience. The editorial team at Routledge has done an excellent job of bringing this book to fruition.

Lastly, I would like to thank my fiancée and frequent collaborator Heather Browning for her unending love, brilliant feedback, and thorough proofreading and copyediting. This book has greatly benefited from innumerable discussions with a philosopher of her calibre.

FOREWORD

This book is a fascinating read for anyone who is interested in questions about consciousness from a biological and psychological perspective. It is also of great appeal to a much broader audience, e.g. those who are interested in the interdisciplinary nature of cognition and consciousness, those who adhere to the importance of philosophy in addressing scientific issues, and those who promote other interdisciplinary endeavours such as the integration of science, arts, and the humanities to address big picture questions about the awareness of being, the questioning of realities, the subjective experience of seeing, feeling, doing, and thinking.

In terms of the interface between philosophy and biological sciences, questions about consciousness abound. What aspects of awareness might we share with other members of the animal kingdom? What might it be like to have thoughts without words and reflect on these metacognitive processes, such as *Mental Time Travel*, the ability to think about how our thoughts change in other times, both past and future, and *Theory of the Mind*, the ability to be aware that others may have both same and different thought processes to our own? What are the evolutionary origins and functions, or in other words how and why have these abilities developed through evolutionary time? How should we consider diverse intelligences within the framework of consciousness? Nagel famously posed the question of "what is it like to be a bat?" and many sceptics would agree that we can never know, for consciousness is a private thought, modulated by personal past experiences and emotions, and hopes and dreams about our subjective thoughts and imaginations of how the future may unfold. Recent developments in the psychology of comparative cognition, in evolutionary biology and in philosophy of mind, however, add important insights and significant progress in this field, and these are the issues which Walter Veit discusses so elegantly and accurately in this book. For example, recent work in comparative cognition over the past couple of decades has provided persuasive evidence that some non-human animals are capable of Mental Time Travel – from corvids (members of the crow family), including the raven pictured on the front cover of the book, to cephalopods (cuttlefish, octopus, and squid). Work in evolutionary biology, in combination with neuroscience and philosophy, has also made a significant impact such as the *Cambridge Declaration on Consciousness* in 2012 and the formation of the first interdisciplinary journal of non-human consciousness *Animal Sentience* in 2015. One should also note the recent work on animal welfare and cognition, and its significance for developing a new UK Animal Welfare (Sentience) Act in 2022, based on the impact of recent developments in animal welfare and cognition, situated within a philosophical framework of the dimensions of animal consciousness.

In the preface, Walter Veit clearly states the book's objective, namely to formulate and develop a biological science of consciousness that focuses on the evolutionary origins, function, and phylogenetic diversity of consciousness and catapult such a framework from the sidelines to the very centre of considerations about the science of consciousness. Walter Veit is exceptionally well placed to achieve this goal, and to integrate these disparate fields, given his background in philosophy and his experience in also working with experts in biology and psychology. The book also provides a historical approach to the field, with its origins, e.g., in the seminal writings of Donald Griffin as to why biologists should be interested in questions of consciousness in non-human animals, not to mention the importance of philosophers such as Dan Dennett, Jonathan Birch, and Peter Godfrey-Smith; evolutionary biologists such as Charles Darwin and Eva Jablonka; and animal behaviourists such as Marian Dawkins and myself, to name just a few of the influences.

Like most scientists studying consciousness, Walter Veit's approach is to treat consciousness as a complex biological phenomenon that has been developed over the vast history of evolutionary time. That said, his approach is novel and exciting because he offers an intriguing strategy, namely to consider animal life histories as well as focusing on the healthy and pathological varieties and gradations of consciousness as a complex phenomenon in nature. The argument is that one should deploy an evolutionary inspired bottom-up approach to investigate whether non-human animals have different dimensions of consciousness and to what degree, as opposed to a top-down investigation of whether evidence of non-human animal awareness meets the criteria for human consciousness. He argues that animals may have subjective awareness at various levels of perception of vision and touch, as well as the evaluation of the hedonic value of experience, and the way in which they may be able to integrate these experiences and explore the motivational trade-offs in decisions now and in the future, and what that might mean for their understanding of the subjective nature of the projection of the self in time. In doing so, he truly integrates the philosophical with the biological and psychological. Why philosophical? Suffice it to say, this is all set within a framework of phenomenological and pathological complexity and how this results in the "Final Crowning Chapter of the Darwinian Revolution".

I shall reveal no more: it's up to the reader to discover and decipher. What I would say is that this is a beautifully written book full of surprising insights, marvellous metaphors, and wonderful titles that inspire and intrigue. I recommend this book in the highest possible terms.

Nicola S. Clayton FRS
Professor of Comparative Cognition,
Department of Psychology,
University of Cambridge, UK

1

A DARWINIAN PHILOSOPHY FOR THE SCIENCE OF CONSCIOUSNESS

1.1 Introduction

This monograph is a philosophical contribution to the emerging science of animal consciousness. It is a science that the prominent American ethologist Donald Griffin tried to establish in the 1970s when he called for a "cognitive ethology" that applies Darwin's lessons to consciousness:

> Most of Darwin's basic ideas about evolution are now generally accepted by scientists, but the notion that there has been evolutionary continuity with respect to conscious experiences is still strongly resisted. Overcoming this resistance may be the final, crowning chapter of the Darwinian revolution.
>
> *Donald Redfield Griffin (1998, p. 14)*

But even a decade after his death, with the "Cambridge Declaration on Consciousness" in 2012 and the formation of the first interdisciplinary journal of non-human consciousness in 2015, aptly titled "Animal Sentience", we remain far from achieving a Darwinian study of the mind. The goal of this book is thus to advance Griffin's vision of the *final, crowning chapter of the Darwinian revolution* by helping the burgeoning field of consciousness research to cast off the chains of a pre-Darwinian view of the mind in both philosophy and science. This will allow us to transition towards a true Darwinian science of consciousness in which the evolutionary origin, function, and phylogenetic diversity of consciousness are moved from the field's periphery to its very centre and enable us to endogenize consciousness into an evolutionary view of life.

In the preface, I have emphasized that there are two senses in which we can ask about the place of consciousness in nature: one concerns the presence and contents of minds in nature and the other the relationship between matter and mind. The former has been called the *problem of other minds*,[1] the latter is the familiar *mind-body problem*; two problems to which answers vary so incredibly widely that many are under the impression that there is no standard according to which we could even begin to sort out which views are likely going to be wrong. It is in this context that one may be forgiven for thinking that the historical question of how consciousness evolved constitutes anything but an additional problem that only further complicates the picture. Yet, it is precisely in the modern twenty-first century theory of evolutionary biology that we find the much-needed standard that the science of consciousness was so desperately lacking.

DOI: 10.4324/9781003321729-1

In order to develop a true biological science of consciousness, we must attend to the (cognitive) ethologist's demand to address the *functionalist question* of what consciousness, in all of its diversity and gradations, does for healthy agents within their normal ecological lifestyles and the natural environments they have evolved in. Accordingly, this book has two objectives: (i) to argue for the need for and possibility of an evolutionary bottom-up approach that addresses the problem of consciousness in terms of the evolutionary origins of a new ecological lifestyle that made consciousness worth having and (ii) to articulate a thesis and beginnings of a theory of the place of consciousness as a complex evolved phenomenon in nature. This thesis can be succinctly summarized as follows.

The Pathological Complexity Thesis:
The function of consciousness is to enable the agent to respond to pathological complexity.

Inspired by Godfrey-Smith's (1996a) *environmental complexity thesis* that sought to establish a link between environmental complexity and the evolution of cognitive complexity, "The function of cognition is to enable the agent to deal with environmental complexity" (p. 3), the pathological complexity thesis is grounded in the idea that health and consciousness are two closely related natural phenomena.[2] Not only will I argue that the origin and function of consciousness lie in the capacity to help complex but vulnerable animals deal with their species-specific health challenge to seek out the beneficial and avoid the pathological, but also that a naturalist understanding of this "biological normativity" requires the development of a Darwinian theory of the organism that will in turn allow us to make sense of organisms as active *agents* and *subjects* including their subjective experience as an integral part of our biological understanding of what makes a bat a bat, a snake a snake, and a healthy bee a healthy bee.[3]

1.1.1 Pathological and Phenomenological Complexity

The pathological complexity thesis is intended as a functionalist alternative to the false dilemma between the two dominant traditions in the philosophy of mind and the science of *human* consciousness, i.e. between strongly externalist representationalist theories of consciousness that overemphasize sensory experience and strongly internalist ones that overemphasize self-awareness as the models for all of the experience.[4] Instead, the pathological complexity thesis seeks to develop an alternative model of consciousness based on a model of *animal sentience.*

Because of the associations of the term "consciousness" with the complexity of the human mind, the term "sentience" – coming from the Latin verb *sentire,* i.e. "to feel" – is often preferred among those with a primary interest in animal consciousness.[5] The term has not received universal endorsement, however, because it is often used ambiguously as (i) a deliberately broad and inclusive concept to refer to all kinds of subjective experiences, (ii) a reference to the most minimal kind of subjective experience found at the evolutionary origins of consciousness, or (iii) the hedonic capacity to feel pleasure or pain. Here, we can avoid these ambiguities because this book will combine all three interpretations. The origins and raison d'être of minimal consciousness or "qualia" lie in hedonic evaluation as "valence" (rating experiences as good, neutral, or bad). Sentience in this evaluative sense is an inherently "interactionist" – or perhaps better, "dynamic" – dimension of consciousness.

Pathological complexity is neither an internalist nor an externalist measure but emerges dynamically from the interaction of organisms and their environments as a measure of the complexity of an organism's life-history strategy and will hence vary with the different "lifestyles" of different animals. It can be understood as the computational complexity of the Darwinian trade-off problem faced by all biological agents as they deal with challenges and opportunities throughout their life histories in order to maximize their fitness. As I shall argue in this book, consciousness evolved during the Cambrian explosion 540 million years ago alongside a new evaluative animal *lifestyle* characteristic of large parts of the metazoan branch of life, as an adaptive response to a computational explosion in just this kind of pathological complexity that made sentience worth having.

Importantly, I use the term "pathological complexity" instead of the equally adequate and perhaps less confusing terms "teleonomic complexity" and "life-history complexity", not because I want to make the argument that organisms with greater life-history complexity are unhealthy, but because I want to emphasize that it is only in understanding life-history trade-offs that we can distinguish healthy from pathological trait variations of traits and that includes variations of consciousness both within and across species, i.e. what I shall term "phenomenological complexity".

Unlike other theories of consciousness that struggle to make testable predictions, the pathological complexity thesis offers us a conjectural empirical framework for the relationship between mind and life by linking properties of phenomenological complexity – such as sensory experience, self-awareness, hedonic feelings, points of view, and mental-time travel – to properties of pathological complexity. A deeper understanding of what makes varieties of consciousness healthy and pathological will be of utmost importance for extending the Darwinian revolution towards consciousness, which is why parts of this chapter will be dedicated to explicating health as a natural phenomenon.

Operationalizing pathological complexity in terms of the number of parameters and constraints in the evolutionary optimization problem studied by state-dependent or state-based behavioural and life-history theory offers us an elegant framework to naturalize health, organisms, and the idea of different ecological lifestyles central to a Darwinian approach of life and mind. It is my hope that the thesis of this book will provide us with a fruitful hypothesis and framework to move us closer towards a comparative science of animal consciousness that can help us to make sense of the place of mind in nature.

1.1.2 Chapter Outline

This chapter is organized as follows. Section 1.2 "Some Preliminary Remarks on Organisms, Health, and Philosophical Method" offers some meta-philosophical reflections about naturalist philosophy and how we can make progress, despite resistance, to understanding notions such as health and consciousness from a Darwinian point of view. Section 1.3 "Lessons from the Darwinian Revolution" defends an extension of the Darwinian program towards animal consciousness by placing it in its historical, methodological, and social context. The history of the Darwinian revolution for biology and psychology will offer a number of important scientific and philosophical lessons and building blocks for the pathological complexity thesis that will accompany us throughout this monograph. Section 1.4 "Carrying Darwinism to Completion" combines the foregoing lessons from the Darwinian revolution with modern state-based behavioural life-history theory to build a theory of health, organisms, and ecological lifestyles grounded in *pathological complexity* that can be used to endogenize consciousness into the Darwinian revolution. Finally, Section 1.5 "Outline of the Book" concludes with an outline for the next five chapters describing how pathological complexity will allow us to make sense of the place of phenomenological complexity in nature.

1.2 Some Preliminary Remarks on Organisms, Health, and Philosophical Method

While my suggestion to link health and consciousness in terms of an association between the biologically normative properties of pathological complexity and the phenomenological complexity of organisms will be intuitive to many animal consciousness researchers focusing on sentience and evaluation, many contemporary philosophers may find it strange. Just as consciousness constitutes perhaps the core problem in the philosophy of mind, the proper definitions of health, pathology, and normal functioning constitute the fundamental problems in the philosophy of medicine. This is plausibly part of the reason why philosophers of mind have been so reluctant to seriously consider evaluation – an inherently normative notion – as the most basic form of consciousness. After all, how could one hope to address one of the biggest problems in philosophy by solving another seemingly unrelated problem in an entirely different field?

An immediate concern should be that it is one thing to aim for progress on one of the major philosophical debates, but it is quite another to undertake the task of making progress on two of its most disputed controversies. Furthermore, the idea that these vexing phenomena share an intimate, but yet, unexplored connection may strike some more traditionally inclined philosophers as a bizarre project. Philosophy in the eyes of those within the tradition of Bertrand Russell's decompositional style of analytic philosophy is engaged with the detailed and narrow, rather than with the general and broad, but this is not the vision of philosophy that I endorse here.

1.2.1 A Darwinian Philosophy of Nature

This monograph follows a particular naturalist style of doing philosophy that is common in the sphere of Australian philosophy of biology and psychology. It is advocated by Antipodean philosophers such as Kim Sterelny, Paul Griffiths, and Peter Godfrey-Smith, but also embodied by the likes of Daniel Dennett and Ruth Millikan, in which the biological sciences – and in particular modern evolutionary theory – become an instrument for the materialist philosopher: "a lens—through which we look at the natural world" (Godfrey-Smith, 2013b, p. 4). Godfrey-Smith (2013b) has called this activity "philosophy of nature" to reflect the older ambitions of the German tradition of *Naturphilosophie* to combine science and philosophy to make sense of the world and our place in it, though without excess romanticism and metaphysical speculation, which nicely describes the project of this book.

The intended task of the philosopher here is to synthesize, rather than to analyse, the products of the sciences in order to construct better theories and models – which brings it in many ways indistinguishably close to the kind of integrative work done by important scientific names, such as Darwin himself. Indeed, this book is very much motivated by the idea that we can endogenize consciousness within modern evolutionary biology by following in the footsteps of Darwin and his followers. But my ambition is not merely scientific; it also has a distinctive philosophical flavour that was once beautifully expressed by Richard Rorty who described philosophy as being in the unique position of providing the only "place in the university where a student can bring any two books from the library and ask what, if anything, they have to do with each other".[6] This comparative and integrative ambition is very much the spirit of this big-picture book on the connection between pathological and phenomenological complexity across the animal branch of life. But before we can even begin to instigate such an investigation, we first need some conceptual grip on the nature of our target phenomena.

1.2.1.1 What Is Health and Consciousness?

One immediate philosophical problem for any biological investigation of consciousness and health is that the terms "consciousness" and "health" are notoriously ill-defined. The cognitive ethologist Frans De Waal (2016), for instance, notes that he prefers "not to make any firm statements about something as poorly defined as consciousness. No one seems to know what it is" (p. 23). Former zookeeper and animal welfare expert turned philosopher, Heather Browning (2020b), similarly expressed scepticism that health reflects "any naturally existing state", instead of a mere cluster of different phenomena (p. 164). If they are right, the pathological complexity thesis seems to rest on shaky ground, built to connect two phenomena that may not even exist.

But the absence of precise definitions for either should not stop us in our tracks. Both terms – as used by the public – may be vague, ambiguous, and resistant to the analytic philosopher's ideal of a conceptual analysis that could provide us with a clear-cut definition. Indeed, if one's goal is to provide a definition of the term that would cover its varied usages, one may be tempted to conclude that we would be better off eliminating their folk concepts altogether.[7] But my goal is not conceptual analysis; it is *conceptual explication* (Carnap, 1950) or as I have called it elsewhere *naturalist conceptual engineering* (Veit & Browning, 2020b). We are trying to capture a phenomenon in nature, for which the neurophilosopher Patricia Churchland (2002) suggests that we should simply rely on common sense to establish "provisional agreement" on a number of "unproblematic examples of consciousness" (p. 133). There is no need to provide a philosophically satisfactory concept of consciousness or health before we can begin to investigate them, any more than we would need to define the concept of koala before we can learn about their enjoyment of eucalyptus leaves.

Scientists repeatedly proceed to investigate phenomena that have so far remained elusive, proving that vagueness need not be an obstacle to scientific inquiry (Neto, 2020). In this naturalist activity, it is ultimately nature, not intuition, that will decide how we should understand consciousness, precisely because as Figdor (2018) argues, we "lack widely accepted theories and models that can organize and articulate the pre-theoretic consciousness-related concepts we are using to guide our initial investigations" (p. 10). Following Churchland (2002), we can confidently reply that we can at least initially "use the same strategy here as we use in the early stages of any science: delineate the paradigmatic cases, and then bootstrap our way up from there" (p. 133).

Paradigmatic cases of consciousness are plenty: pain, pleasure, smell, vision, taste, a sense of one's body, and memories, alongside a whole other range of subjective experiences. Similarly, we have some intuitive grasp of health and pathology in humans and animals alike, such as diseases, broken bones, lesions, parasites, burns, poisons, maladaptive behaviour, and other "biological wrongs" – even if we have struggled to derive something like a folk theory of health. So it is perhaps unsurprising that we can also intuitively distinguish healthy subjective experiences from unhealthy ones such as major depressive disorder, anxiety disorder, aphantasia, synesthesia, autism, schizophrenia, prosopagnosia, chronic pain, and many more. Yet, many philosophers of medicine would *deny* that health is a natural phenomenon, thus perhaps providing an explanation for why philosophers have given so little attention to the search for the origins of consciousness in the normative notion of evaluation.

1.2.2 Resistance to Naturalism in the Philosophy of Medicine

Despite naturalist views being discussed by philosophers of medicine, their assessment is largely negative, and most within the field now maintain that health reflects personal evaluations or the values of society at large; a consensus that health is primarily a *normative* concept, rather than "only" an objective *biological* property of organisms.[8]

Such a view may have been the dominant one ever since the French historian of science and first modern philosopher of medicine, Georges Canguilhem (1991), argued in his influential treatise *The Normal and the Pathological* that "[t]here is no objective pathology. Structures or behaviors can be objectively described but they cannot be called [']pathological['] on the strength of some purely objective criterion" (p. 226). Others like Lennart Nordenfelt (1995), who emphasize the concept of agency, have argued that health cannot be understood in a reductionist naturalist way and instead requires a more holistic conception, where it is understood as the ability to achieve one's vital goals. Phenomenologists such as Havi Carel (2007) have similarly argued that the "experience of illness cannot be captured within a naturalistic view" (p. 95). Such strong assertions against the very *possibility* of a naturalist account are surely premature and yet can be found throughout the literature, effectively making naturalism a "boogeyman" of the field. Rarely has there been a philosophical debate in which naturalism has been so forcefully and unceremoniously dismissed.

This anti-naturalist consensus in the field can be usefully summarized as an appeal to the "irreducibility" of (i) the normativity of health and disease, (ii) the loss of agency in health and disease, and (iii) the phenomenology or subjective experience of health and disease. But who is to deny that these features can be part of a naturalist account of health and disease? Naturalist philosophers have long worked on attempts to make these notions of normativity, agency, and phenomenal experience safe for naturalism.[9] What all of these anti-naturalists curiously share, though not necessarily all other philosophers of medicine opposed to a naturalist view of health, is an emphasis on *subjectivity*. Akin to those who view naturalist explanations of consciousness as deeply problematic, they argue that the very idea of a naturalist account of health and disease is mistaken. They hold that one cannot account for health and disease from the objective third-person perspective of science, since they are phenomena at the level of a *subject*, not an *object*, and science cannot account for the former – a view familiar from so-called naysayers who assert a scientific account of consciousness to be impossible.[10]

This way of thinking about naturalism, however, is highly problematic. Subjects aren't some mysterious entities inaccessible to science: they are an evolutionary product and also include non-human animals. But the possibility of a Darwinian reconciliation between a view of health as a property of the organism as an "object" and of the organism as a "subject" has been given scant attention, precisely because *non-human health* has been less than an afterthought in this debate (see Matthewson & Griffiths, 2017).

As I will argue in this chapter, not only health and pathology are perfectly naturalistic concepts but they also play a key role in evolutionary biology and will help us to extend the Darwinian revolution to include consciousness.

1.3 Lessons from the Darwinian Revolution

The first chapter of this book is titled "A Darwinian Philosophy for the Science of Consciousness" precisely because the so-called emergence in the 1990s of a science of consciousness has at best been a science of *human* consciousness. From a naturalist perspective, we can only truly claim to have established a Darwinian science of consciousness once we study consciousness as a natural, rather than a human, phenomenon – and this must include all sentient animals. Unfortunately, consciousness appears to be one of the last biological phenomena that we have failed to integrate into the Darwinian revolution.

As I noted in the preface, this project has been burdened with unfortunate epithets such as "just-so stories", demonstrating a resistance to the possibility of an adaptationist evolutionary process explanation of how the mind gradually came into existence. Lewontin (1998) himself, who has been one of the fiercest opponents of adaptationist explanations in biology,[11] contributed to this scepticism when he argued that we know next to nothing about the evolution of the human mind and probably never will. Such a pessimistic attitude is certainly not entirely unfounded, since consciousness appears to leave no fossil trace, making it seemingly impossible to trace its phylogenetic origins and reconstruct its raison d'être through a historical narrative explanation.

In this, however, consciousness is not alone – sharing a fate with a wide range of other complex biological phenomena that people thought could not be explained in Darwinian terms. Most notably of these is behaviour, which has been firmly integrated into modern evolutionary biology since the ethologists endogenized it within Darwin's explanatory framework. Paying close attention to the origins of the Darwinian paradigm and its extension to behaviour will provide a number of useful lessons for a *cognitive ethology*, which likewise endogenizes consciousness in a Darwinian view of life.

1.3.1 Darwinism and Teleonomy

In trying to provide a Darwinian account of consciousness, we have to clarify what we mean by such a project. Above, I noted that the pathological complexity thesis rests on the Darwinian idea of a functionalist alternative to a false dilemma between externalist and internalist approaches to consciousness. Internalist explanations seek to explain features of a system in virtue of other features of that system – of processes, structures, organization, and development *within* it, rather than outside of it. Externalist explanations, on the other hand, aim to explain features of the system by recourse to the external, i.e. the environment – Godfrey-Smith (1996a) calls them "outside-in" explanations (p. 30). This distinction is not only relevant for categorizing different views of the mind, but also of life itself, since many treat Darwinism (mistakenly) as an externalist program.

Lewontin has stated the alleged link between Darwinism and externalism perhaps the most forcefully, arguing that the success of the Darwinian project was due to its disentangling of internal and external forces that have previously been inseparable (see also Lewontin & Levins, 1997). Darwin broke with what Lewontin called *transformational* theories of the past, such as Lamarck's (1984) theory of evolution that postulated change to individuals within their life histories arising from "subjective" or what we may want to call "internal" forces, such as will and striving. The Darwinian theory of the organism made it the "*object*, not the subject, of evolutionary forces" such as natural selection and random drift that are "autonomous and alienated from the organism as a whole" (1985, p. 85). To complete the Darwinian revolution, however, Lewontin maintained that the internal forces – the subject-side of organisms – must be reintroduced:

> Darwinism cannot be carried to completion unless the organism is reintegrated with the inner and outer forces, of which it is both the subject and the object.
>
> *Richard C. Lewontin in Levins and Lewontin (1985, p. 106)*

By this, Lewontin did not mean subjective experience, but rather how organisms as agents actively "participate" in their evolutionary path and "construct" their environments, as an alternative to a traditional adaptationist view of life. These notions of agency and construction have been highly influential in modern attacks on Darwinism (Ho & Saunders, 1979; Laland et al., 2014; Müller, 2017; Noble, 2015), but I am not here interested in the conceptual role of organisms as subjects for challenging the theoretical modelling of evolution. My interest lies in subjects as an evolutionary product to allow us to make sense of the evolution of subjective experience. As Godfrey-Smith (2017c) notes in his discussion of Lewontin, not only subjects are a cause of evolutionary change, but they are also its product.

In advancing a gradualist view of the evolution of consciousness, theoretically less loaded terms like "agency" and "subjectivity" are useful for thinking about organisms as being more or less subject-like; they "can realize subjectivity to a greater or lesser degree" (Godfrey-Smith, 2017c, p. 1). While subjectivity may appear similarly as elusive as consciousness, it does not similarly suffer from an overabundance of theoretical frameworks. We can, as Godfrey-Smith (2019a) argues, use Lewontin's distinction between objects and subjects to bridge the gap between matter and mind: "[t]he history of life includes the history of subjectivity, and subjective experience is the experience *of a subject*" (p. 2). And in doing so, we may be able to carry the Darwinian revolution to its completion.

Unlike Lewontin, however, I do not see a conflict between adaptationism and an explication of the subject-side of organisms. As this book hopes to demonstrate, it is precisely with a Darwinian view of organisms that we will be able to make sense of "subjectivity". This does not mean that we can't recognize that evolutionary biology has been dominated by externalist modes of explanation, with features of the organism being explained in terms of their adaptive fit to their external environment (Godfrey-Smith, 1996a; Walsh, 2015). Evolutionary biologists readily admit that "[t]he suspicion of internal causes in the dominant neo-Darwinian culture ran so deep that every internalist idea, no matter how reasonable, was treated as an appeal to vitalism" (Stoltzfus, 2019, p. 46). But we should distinguish the idealization choices made by *some* modellers, from a deeper commitment to the necessity of an externalist view of adaptations. After all, there is plenty of modelling work done by evolutionary biologists that can be seen as "internalist", such as the study of game-theoretic dynamics that emerge from the structure of a population rather than its external environment (Sterelny, 1997, p. 556). Indeed, it is a mistake to think of adaptationism and externalism as a one package deal. As I shall argue, we can straightforwardly follow Sterelny's (1997) suggestion to decouple adaptationism from externalism and consider the two separately.

Many of the arguments against adaptationism are really arguments against its externalist versions that use a so-called lock and key model of the adaptation between organisms and their environments, a criticism that need not apply to other versions. Modern evolutionary biology recognizes plenty of feedback between organisms and the species-specific environments in which natural selection takes place, such as Brandon's (1990) notions of "selective environments" and "ecological environments", which can be distinguished from an organism-neutral externalist view of the environment. The external features that *matter* to the evolutionary trajectory of the organism are themselves causally dependent on the organism. No longer do modern evolutionary biologists see adaptations in the externalist design-sense of a natural theologian such as Paley (1802), who argued that animals are a proof of God's design plan, with species being fitted to pre-existing external niches.

As with many scientific concepts, the concept of adaptation came to be redefined – or rather explicated – in a naturalistically unproblematic sense referring to whatever is produced by natural selection, even if such "design" appears inefficient and wasteful (Griffiths & Gray, 2001, p. 209). Much of the opposition from "Neo-Darwinians" to Gould's and Lewontin's criticism of "adaptation" was based on a mismatch between a usage of that term in its original pre-Darwinian sense and its modern explication, which already included at least some of the features of feedback between organisms and their environments that were alleged to be lacking in the modern neo-Darwinian view of life. Instead of seeing Darwinism as an externalist theory of organismal traits that replaced previous vitalist and romanticist modes of thinking that were confused between internal and external forces, we should see it as a *teleonomic* rejection of a false dilemma between internalist theories such as Lamarck's and a strongly externalist view of organisms being designed by a benevolent God to fit their environments, by providing us with an inherently "dynamic" or "interactionist" picture of the living world.

By "teleonomic" I am employing Pittendrigh's (1958) coinage of the term, as a naturalistically unproblematic Darwinian replacement for older and mistaken teleological notions about the purposefulness, design, and normativity of life, which is why I noted above that my notion of "pathological complexity" could alternatively have been called "teleonomic complexity". By understanding organisms as goal-directed systems or Darwinian *agents* evolved to maximize their fitness, our understanding of health, just like our understanding of adaptation and design, will come to be transformed. As I shall argue in this book, we can build a theory of the organism as both an object and a *subject* with the tools of modern state-based and behavioural life-history theory, which does not – as Lewontin objected to – treat organisms as machines with mosaic-like traits, but rather as agents having to deal with integrated bundles of trade-offs in organismal design. It is precisely this teleonomic theory that will bring out the subject-side of organisms. With this, let us now turn to Darwin's own speculations about the evolution of mind.

1.3.2 Early Darwinian Views of Mind

As the approach in this monograph is inspired by Darwin, it will be hardly surprising that Darwin himself rejected a dualist view of the mind, both in a metaphysical sense and in the phylogenetic sense of a sharp dividing line between us and other animals. Following the success of his 1859 book *On the Origin of Species*, he published two further very influential books in which he sought to defend a continuity view between us and other non-human animals. In his *The Descent of Man, and Selection in Relation to Sex*, Darwin (1871) argued that "the lower animals, like man, manifestly feel pleasure and pain, happiness and misery" (p. 39) and that "there is no fundamental difference between man and the higher mammals in their mental faculties" (p. 35). And in his *The Expression of the Emotions in Man and Animals*, Darwin (1872) went on to vastly expand his hypotheses on the evolution of mind, in particular the emotions.

But while Darwin urged us to think about the mind in terms of evolutionary continuity, he deliberately avoided public speculation on the very origin of mind and life, noting: "I must premise that I have nothing to do with the origin of the primary mental powers, any more than I have with that of life itself" (1859, p. 207). Yet, it is clear that the evolution of consciousness sincerely troubled him and some of his early notes revealingly contained the questions: "How does consciousness commence?" and "Where pain & pleasure is felt where must be consciousness???", suggesting that even Darwin speculated about the origins of sentience.[12] Later, he repeated his resistance to explaining the origins of life and mind as problems that ought to concern us *now*: "In what manner the mental powers were first developed in the lowest organisms, is as hopeless an enquiry as how life itself first originated. These are problems for the distant future, if they are ever to be solved by man" (1871, p. 36). But as with the origins of life, early Darwinists were immediately spurred on to think about the origin of mind.

Indeed, the idea of thinking about the mind as a product of evolutionary forces immediately influenced important figures such as Herbert Spencer, Thomas Henry Huxley, Ernst Heinrich Philipp August Haeckel, George John Romanes, William James, Conway Lloyd Morgan, James Mark Baldwin, and John Dewey, who all substantially contributed to an early evolutionary understanding of the mind.[13] What we saw in the decades after Darwin, Ginsburg and Jablonka (2019) note, was that "all psychologists, philosophers, and biologists who considered mental evolution and the evolutionary origins of mentality explained it in terms of natural selection" (p. 71).

Indeed, it was common at this time to think that the mysteries of the mind could be unveiled by viewing them through an evolutionary lens.

Spencer, for instance, insisted that "[i]f the doctrine of Evolution is true, the inevitable implication is that Mind can be understood only by observing how Mind is evolved" (1870, p. 291). Moreover, both Dewey and Spencer endorsed a continuity thesis between life and mind that influenced my own: the mind is seen as the natural consequence of the evolution of complexity.[14] Furthermore, Romanes speculated in some detail that pleasure and pain may be the key to understanding the place of consciousness in nature:

> Possibly, however—and as a mere matter of speculation, the possibility is worth stating—in whatever way the inconceivable connection between Body and Mind came to be established, the primary cause of its establishment, or of the *dawn of subjectivity*, may have been this very need of inducing organisms to avoid the deleterious, and to seek the beneficial; the raison d'être of Consciousness may have been that of supplying the condition to the *feeling of Pleasure and Pain.*
>
> *George John Romanes (1883, p. 111) [italics added for emphasis]*

Evolutionary thinking naturally lends itself towards a view in which sentience constitutes the origin of consciousness. That organisms would evolve to value states and behaviours that increase their own fitness and avoid those that are detrimental to their health appears not at all mysterious from a Darwinian point of view.

Unfortunately, discussions of consciousness and its evolution, in both humans and non-human animals, went out of fashion in the early twentieth century. This was largely as a result of the rise of the behaviourist program coming from Watson and the more radical behaviourism of Skinner, who turned *American* psychology into the study of mere behaviour, banning consciousness from science. But while their official doctrine has been all but abolished, their influence in the study of consciousness remains alive and well. In the following, we will take a closer look at the rise of behaviourism, classical ethology, and Griffin's eventual call for a cognitive ethology in order to understand what it means to take a truly Darwinian approach to life and mind.

1.3.3 Jamesian Psychology and the Rise of Behaviourism

To understand the rise of behaviourism, one must understand the status of psychology at the beginning of the nineteenth century. One name that has perhaps influenced the science of consciousness more than any other is that of the aforementioned American philosopher and psychologist William James.

James is often credited for turning psychology into a discipline independent from philosophy with his 1890 textbook *The Principles of Psychology*, an achievement that made him the so-called father of American psychology in the eyes of many. In the early development of psychology as a science, consciousness played an important role, so it should hardly be surprising that James is also praised as the "father of modern consciousness studies" (Ginsburg & Jablonka, 2019, p. 41). Unfortunately, James had little to say about animal consciousness or its evolutionary origins, despite his interesting speculations about consciousness as the emergence of a new kind of evaluative agency and his emphasis on a functionalist view of the mind. His explanatory target was ultimately human consciousness, which he believed was undeniable, almost unique in kind, and could best be studied through the method of personal introspection.

The focus of psychology on consciousness, however, quickly came to be questioned. With the further development and success of psychological experiments, appeals to subjective states were less and less seen as "necessary to justify the value of experimental research" (Burghardt, 1985, p. 914). The behaviourist program that tried to banish all mental concepts from psychology, and turn the science of the mind into a science of behaviour, can be seen as a natural outcome of this trend with functionalism coming to be abandoned. However, this did not mean that the behaviourists were anti-Darwinian, at least not initially. Indeed, unlike James who centred psychology around human consciousness, the behaviourists positively emphasized the importance of studying non-human animals due to their evolutionary continuity with us.

It is unfortunate that Watson, who is usually credited as being the father of the behaviourist movement, is often demonized and misdescribed. When attention is given to his early work, "presentations are usually brief and frequently contain a variety of errors" (Todd & Morris, 1986, p. 71). Rather than treating behaviour as a black box, Watson showed a keen interest in the neurophysiology of animals, dissecting them with great care and experimental detail. However, after a decade of rigorous and methodologically diverse work on animal behaviour, Watson was ultimately fed up with having to justify the value of his research, after being repeatedly faced with the sceptical question of what his work could possibly teach us about human consciousness.

This should immediately remind us of the question not uncommon in twenty-first century *human* consciousness science and the philosophy of mind, regarding what the bearing could possibly be of work on animal consciousness. Watson's response to his detractors could hardly have been more Darwinian. In his 1913 paper "Psychology as the Behaviorist Views It" – the founding manifesto of the behaviourist tradition that was meant to put these critics to rest – Watson explicitly defended the Darwinian view that there is "no dividing line between man and brute" (p. 158). To understand behaviour as a natural, rather than a human, phenomenon, he maintained was how we could only truly advance a science of behaviour as a natural phenomenon. Those inspired by Jamesian psychology, he harshly accused of being stuck in a pre-Darwinian mindset:

> [T]o make consciousness, *as the human being knows it*, the center of reference of all behavior, forces us into a situation similar to that which existed in biology in Darwin's time.
>
> *John B. Watson (1913, p. 124) [italics added for emphasis]*

To understand the phenomenon of life, biologists readily recognized that an exclusive look at humans would lead to a biased picture, if not because of its complexity then because of the appeal to thinking of the human body plan as "perfect" or "higher" than other species. What was needed to truly revolutionize our understanding of life as a natural phenomenon was an evolutionary approach based on phylogeny, the comparative method, and sound ecological thinking. Yet, early work in evolutionary biology was initially held back by its focus on the question of human descent.

This *starting point* was perhaps not surprising in a historical sense, since a sharp dividing line between humans and the rest of nature was considered to be the greatest challenge to Darwin's theory of natural selection. An assumption of human uniqueness had to be overcome. After the continuity between humans and apes was settled, biologists were finally able to put humans in their place in nature, i.e. one among many species; man was dethroned. In trying to understand biological phenomena, biologists would henceforth use the comparative method – gathering evidence from many different species of animals and plants alike to learn general lessons about life. But from the perspective *of* Darwinism as a research program that placed us alongside rather than above all other life forms, this early focus on humans must have seemed strange. As Watson (1913) put it: "Man ceased to be the center of reference" (p. 125).

In arguing against psychology as the "science of the phenomenon of consciousness", Watson provided us with Darwinian arguments that very well apply against the top-down human-centric focus of the so-called science of consciousness of today. But before we turn to Griffin's cognitive ethology as an attempt to develop a bottom-up biological study of the mind, let us first look at the classical ethologists in order to understand how they extended the Darwinian revolution towards behaviour.

1.3.4 Ethology, Health, and the Darwinization of Behaviour

In their introduction to the philosophy of biology, Sterelny and Griffiths (1999) define ethology as "the study of animal behavior under its normal ecological conditions (as opposed to unusual laboratory conditions) and from an evolutionary perspective" (p. 385). And this is certainly how many now think about it, as a tradition that was in opposition to the lack of ecological and evolutionary thinking shown by the behaviourists, and one that has now largely been superseded by behavioural ecology. But there was a more philosophical conviction that motivated its founders, one of a teleonomic view of life, and this has largely gone unnoticed.

When one hears the term "ethology", inevitably the names Konrad Lorenz, Nikolaas Tinbergen, and Karl von Frisch come to mind as the joint receivers of a Nobel Prize in Physiology or Medicine in 1973 for their involvement in the establishment of ethology and their discoveries of "organization and elicitation of individual and social behaviour patterns" (Nobel Prize Outreach, 2021). Famously, Lorenz studied the imprinting behaviour of greylag geese, who show an innate instinct to bond with the first moving entity they encounter, whereas von Frisch was one of the first to study the "waggle dance" used for communication by bees. Tinbergen, who is more well-known than the others primarily due to his involvement in the spread of ethology through the Anglosphere, spent much of his time studying so-called fixed action patterns such as the egg rolling of the greylag goose (see Beer, 2020). But it is not their work on instincts and other proto-cognitive capacities that is of relevance to this book. Neither am I interested in their philosophical objections to the study of subjective experience. The reason we look here at the (classical) ethologists is the same reason we looked at Watson's motivation for the behaviourist manifesto: i.e., to emphasize a Darwinian principle that motivated the origins of their approach.

Both the ethologists and the behaviourists wanted to establish an objective science of behaviour in which we rely on a bottom-up approach that emphasizes the study of simple behaviours in order to understand more complex ones. But the ethologists hardly saw the behaviourists as Darwinians at all. This is ironic, considering that both the (early) behaviourists and ethologists used Darwin to motivate their approach. However, we can readily resolve this puzzle. Whereas the behaviourists emphasized the alleged externalist explanatory style of Darwin's theory of natural selection, ethologists emphasized the theory itself, with its emphasis on function, survival value, and evolutionary phylogeny as sources of mechanisms to deal with the environments faced by organisms. This teleonomic perspective is nicely drawn out in a press release from the Karolinska Institute, which announced the Nobel Prize for the founders of ethology, and described their approach in a very Lorenzian manner as a Darwinian way out of a dilemma between the behaviourist's externalism and the vitalist's insistence on internalist forces:

> During the first decades of this century research concerning animal behaviour was on its way to be stuck in a blind alley. The vitalists believed in the instincts as mystical, wise and inexplicable forces inherent in the organism, governing the behaviour of the individual. On the other hand reflexologists interpreted behaviour in an one-side mechanical way, and behaviourists were preoccupied with learning as an explanation of all behavioural

> variations. The way out of this dilemma was indicated by investigators who focused on the *survival value* of various behaviour patterns in their studies of species differences. Behaviour patterns become explicable when interpreted as the result of *natural selection*, analogous with anatomical and physiological characteristics.
>
> *Nobel Prize Outreach (2021) [italics added for emphasis]*

Following the end of the First World War, American psychologists came to treat appeals to instincts as unscientific explanations (Griffiths, 2008). Instincts – in the sense of *unlearned* responses – were seen with unease, due to their teleological and internalist nature, their inability to yield themselves to physiological investigation and thus likewise causal explanation, in addition to their being tied up with mistaken purposive and vitalist conceptions of life (Dunlap, 1919; Griffiths, 2008; Kuo, 1921; Tolman, 1923). Whereas the likes of Darwin and James strongly endorsed the idea of instincts, with the externalist turn of the behaviourists the notion came to be seen as unscientific. Lorenz, however, resisted this response as something that went too far in the opposite direction, maintaining that "it is hardly an exaggeration to say that the large and immeasurably fertile field which innate behaviour offers to analytic research was left unploughed because it lay, as no man's land, between the two fronts of the antagonistic opinions of vitalists and mechanists" (1950, p. 232). Indeed, Lorenz simply tried to do for behaviour what Darwin's naturalism had previously achieved for a similar false dilemma between vitalists and mechanists on the nature of life and organismic activity, i.e. to emphasize the teleonomic nature of animals:

> [B]oth mechanists and vitalists were incurably inhibited by quite specific conceptual errors and prejudices that were magnified by their clash of opinion. This prevented them from initiating research into animal and human behavior at the point where it should have begun, namely with straightforward, unprejudiced *observation* of healthy animals living under normal conditions. They were quite incapable of seeing behavior for what it is, that is, as an extremely complex, organic *systemic entity* consisting of quite different components; one which, like *any* organic system, owes its particular constitution to a quite specific historical process of development.
>
> *Konrad Lorenz (1997, p. 214) [italics in original]*

To understand ethology as the mere opposition to exclusive laboratory work would be to miss out on this most important observation: ethology was intended as a teleonomic science (see also Thompson, 1986a, 1986b). In claiming to study learning in healthy organisms, the behaviourists did not recognize that it made no sense to think of health outside of the ecological context animals evolved in. As Lorenz (1981), who himself earned a Doctor of Medicine, put it: "the pathologic can be defined only by having recourse to ecological concepts" (Lorenz, 1981, p. 57).

Unfortunately, this philosophical insight of the ethological tradition has come to be neglected, next to the more popular methodological slogan of studying animals in the wild, despite the fact that it was precisely their teleonomic reasoning that made them emphasize the importance of studying the lifestyles of healthy animals in their natural environments. This is why Lorenz praised Oskar Heinroth as the real founder of ethology as the comparative study of behaviour:

> Accordingly, a finely developed sensitivity to the delicate and *often diffuse* boundary between the not-quite-normal and the already pathological is perhaps the most important talent that a scientific animal keeper must possess! On the other hand, however, the keeping of animals in our sense is an apprenticeship that renders one's feel for the pathological

> just as acute as actual training in a medical clinic. It is surely no coincidence that Heinroth, the outstanding master of animal keeping as a scientific method, was, like the writer of this text, initially trained as a *medical doctor.*
>
> *Konrad Lorenz (1997, p. 229) [italics in original]*

While the research of both Lorenz and Heinroth was admittedly often only conducted under quasi-natural circumstances, their work was informed by life-history observations of healthy and pathological animal behaviour in its natural environment. This combination of observation and adaptationist thinking was inspired by the descriptive natural history activity of Darwin and the taxonomic activities that made a distinction between healthy and pathological specimens in order to describe what a "normal" individual ought to look like. This was how the Darwinian revolution began, and this was thus how Lorenz thought a biological study of behaviour must get off the ground. It must be able to distinguish healthy from pathological behaviour, just as physiology and taxonomy have distinguished healthy from unhealthy phenotypes. This is why he emphasized early on that behaviour could be treated in just the same way as any other adaptive phenotype:

> What behaviorists exclude from the narrow circle of their interest is not only other learning processes, but simply everything that is not contained in the process of learning by reinforcement-and this neglected remainder is *neither more nor less than the whole of the remaining organism*! [...] What remains uninvestigated is all that makes an octopus an octopus, a pigeon a pigeon, a rat a rat, or a man a man, and, most important of all, what makes a healthy man a healthy man, and an unhealthy man a patient.
>
> *Konrad Lorenz (1981, p. 71) [italics added for emphasis]*

This is the basic motivation behind the ethologist's Darwinization of behaviour and it is unfortunately a lesson the science of consciousness has not yet learnt. If we are interested in making progress on the problems of consciousness within the next century, we must follow the ethologists' dictum to distinguish healthy from pathological variations of consciousness by thinking about their survival value in an ecological context. With these important lessons regarding the importance of health and the possibility of a teleonomic alternative to the dilemma between externalist and internalist explanatory schemes, let us now turn to Griffin's call for a *cognitive* ethology.

1.3.5 Donald Griffin's Call for a Cognitive Ethology

At the beginning of this chapter, I placed an obligatory quote by Donald Griffin who, beginning in the 1970s, tried to break the behaviourist's hold on the study of animal minds. I say obligatory because Griffin, in the last three decades of his life, has done more than any other scientist to re-establish and give credibility to a Darwinian study of consciousness. Admittedly, Griffin would not have had this impact if he hadn't discovered bat echolocation together with fellow Harvard undergraduate Robert Galambos in 1938. When Thomas Nagel visited his institution from 1973 to 1974 and wrote up his famous paper "What Is It Like to Be a Bat?", directly drawing on Griffin's discovery, Griffin felt immediately compelled to write up his thoughts on the possibility of a study of animal consciousness, resulting in his 1976 book with the fitting title *The Question of Animal Awareness: Evolutionary Continuity of Mental Experience*. Against Nagel, who denied the very possibility of learning about the subjective mental lives of other animals, Griffin (1976) argued that we could and ought to have a naturalistic investigation of the minds of other animals using the tried and tested ethological methods of naturalist observation in the wild, controlled experiments, and sound evolutionary reasoning – calling this thoroughly Darwinian discipline of animal minds "cognitive ethology".

But while Griffin managed to break the taboo against the scientific investigation of animal minds, cognitive ethology unfortunately never took off as a significant research program on its own.[15] Worse, the term became for many a dirty word, associated with unscientific speculation, mere anecdotal evidence, and the *sin* of anthropomorphism (Allen & Trestman, 2017). Here, it is important not to confuse "cognitive ethology" with the general study of animal cognition (Allen, 2004). After a long period of struggles, the subject of animal cognition finally received its own journal with the establishment in 1998 of *Animal Cognition* as an interdisciplinary hub for work on animal minds. However, cognition was here merely defined as the information processing in the brain, leaving out the subjective side of the mind. But this was not what Griffin had in mind when he criticized the remnants of the behaviourist program in the 1970s. His cognitive ethology was meant to investigate all aspects of the mind, with a special emphasis on its subjective character. He wanted to think about what the mind does *for* animals in nature. What are the problems animals solve in their natural ecological niches? As Ristau (1992) put it, the "cognitive ethologist focuses on problems faced in an animal's *natural world*; which is of course the difference between classical ethology and comparative psychology come again" (p. 125). One should therefore not mistake his call for a cognitive ethology as a mere attempt to extend the cognitive sciences to non-human animals.

Unfortunately, it took a decade after his death for such an interdisciplinary field of animal consciousness research to emerge. One hundred years after the behaviourist revolution, we are finally beginning to see this chapter being written. As Birch, Schnell, and Clayton (2020) [henceforth Birch et al., 2020] put it in a recent paper in *Trends in Cognitive Sciences*: "[I]n the past 5 years, an interdisciplinary community of animal consciousness researchers, drawn from neuroscience, evolutionary biology, comparative psychology, animal welfare science, and philosophy, has begun to coalesce around" the questions of: which animals have consciousness and what their subjective experiences are like? (p. 789).

The science of (human) consciousness has finally moved on from seemingly endless debates on whether or which non-human animals are conscious. We are finally rigorously studying the contents of non-human animal experiences. Something of a scientific consensus has been reached that at least some non-human animals are conscious, thus opening the door for a truly Darwinian science of consciousness, such as Griffin envisioned. As Allen and Trestman (2017) put it in their Stanford Encyclopedia article on animal consciousness, we shall be forever grateful for Griffin's "crucial role in reintroducing explicit discussions of consciousness to the science of animal behavior and cognition, hence paving the way for modern investigations of the distribution and evolutionary origins of consciousness". Yet, there remain major conceptual and methodological disagreements on how we should study animal consciousness.[16] As I hope to show in this book, we can make progress towards a genuine science of animal consciousness by following Griffin's call to study consciousness in the spirit of a cognitive ethologist, i.e. as an adaptive and evolved phenomenon in nature.

1.4 Carrying Darwinism to Completion

Throughout the last section, my goal has been to emphasize what it means to take a naturalist approach to the place of consciousness in nature. By this I do not (only) mean the now common understanding of naturalistic thinking as the need for a strong continuity between science and philosophy, but rather the older meaning, of a *natural history* approach that begins with the careful observation of the diversity of organisms in the wild. It is in this intellectual tradition of taxonomic classifications and the mapping out of the life histories of different organisms, rather than in laboratory experiments, that the Darwinian revolution began and changed biology forever. In order to understand organisms, natural historians began by meticulously describing the living world. This led them to appreciate the distinction between healthy and pathological variations, and eventually to a teleonomic theory to make sense of this biological normativity in a naturalistically unproblematic manner.

Hence, it ought not to be all that surprising that the natural historian Alfred Russell Wallace came up with the idea of evolution by natural selection independently from Darwin, while suffering from a fit of malarial fever on his explorations (Meyer, 1895). Convinced of the common origin of species but lacking a process explanation, the fragility of his own health and vigour made him realize that varieties within populations would lead to differences in the adaptive fit of organisms to their environments and thus change species in a "struggle for existence" (Darwin & Wallace, 1858, p. 54).[17] Furthermore, just like Heinroth, Lorenz, and von Frisch, Darwin himself pursued a degree in medicine. While Darwin even grew up in a family of physicians, he eventually gave up on the pursuit of medicine due to his distaste for operations and his greater obsession with natural history (Antolin, 2011). Nevertheless, it is hard to believe that an appreciation for the healthy and pathological varieties of organisms did not leave an impact on his teleonomic thinking about the appearance of "design" in nature. After all, Darwin (1859) himself maintained that we can comfort ourselves that despite the great destruction in the struggle for life, "the vigorous, the healthy, and the happy survive and multiply" (p. 29). Unfortunately, very little attention has been given to the deep link between health, evolutionary theory, and natural history outside of the evolutionary medicine movement and parts of ethology.

In order to understand healthy organisms in their natural environments, ethologists explicated the life histories of organisms through observation and the creation of so-called ethograms, which were intended as an objective description of an organism's behavioural repertoire during their lifetime. Just as the taxonomists and natural historians prior to Darwin distinguished normal from pathological organisms, ethologists aimed to begin a Darwinian study of behaviour by understanding its healthy and pathological variations. It will therefore hardly be surprising that ethograms are still a common tool for both veterinarians and animal welfare scientists for detecting pathological or abnormal behaviour in animals such as tail-biting (Brunberg et al., 2011) and feather pecking (Rodenburg et al., 2003). Yet, this is not a theory of health; ethograms only constitute a list of healthy and pathological behaviours.

It may thus not be surprising that many practitioners such as medical professionals, veterinarians, and animal welfare specialists such as Browning confidently treat health as something like a mere construct rather than a genuine integrated phenomenon in nature. They grant that we can measure parasite load, lack of nutrients, cancer, and particular diseases, among other vulnerabilities and dysfunctions, and that we could rank an animal's "health" with respect to each, but they would deny that there are some objective means of integrating them meaningfully into a single state. Something more is needed to naturalize health as a genuine whole-organism phenomenon in nature and for this we unsurprisingly require a theory of the organism.

This insight has notably already been made by Darwin's grandfather Erasmus Darwin, who was a very influential physician in a family of medical doctors and who had an acute interest in natural history. At this time, natural history was often studied as a resource for medicine and he even anticipated some unironically "proto-Darwinian" ideas in his treatise *Zoonomia*, a book on natural history that set out "to reduce the facts belonging to ANIMAL LIFE into classes, orders, genera, and species; and by comparing them with each other, to unravel the theory of diseases" (Darwin, 1794, p. 1).[18] Such a theory was ultimately derived by Darwin, achieving his grandfather's goal to provide a "theory founded upon nature, that should bind together the scattered facts of medical knowledge, and converge into one point of view the laws of organic life" (Darwin, 1794, p. 1).[19]

While Darwin himself had little to say about health and pathology, his teleonomic theory of evolution by natural selection was later used by the ethologists to likewise synthesize our scattered knowledge about normal and pathological behaviour into the very same evolutionary framework, casting their Nobel Prize for Physiology *or* Medicine in a particularly interesting light. Our folk understanding of "health" as a biological phenomenon is revised in the light of evolutionary theory, just like that of "design". To advance the goal of a true biological science of consciousness, we must – similarly to the ethologists – make use of Darwin's theoretical framework to synthesize our scattered knowledge about healthy and pathological cases of phenomenological complexity across the tree of life to integrate consciousness into the final, crowning chapter of the Darwinian revolution and complete our understanding of diverse organisms, such as electrosensing platypuses, echolocating bats, and infrared-sensing snakes. Or to borrow the words of Lewontin: in order to carry Darwinism to completion, we must understand the organism as both an object and a subject.

Since I've repeatedly stated that state-based behavioural and life-history theory is the key theoretical resource for this task as the theory of teleonomic agency, I am now going to explain this theory in more detail and connect it to the foregoing lessons about the Darwinian revolution.

1.4.1 A State-Based Behavioural and Life-History Theory of the Organism

The twenty-first century equivalent of the ethologists' attempt at building ethograms of organisms is modern state-based behavioural and life-history theory. All organisms go through a life cycle in which they are born, take in nutrients, grow, reproduce (whether sexually or asexually), and ultimately die (i.e. if death hasn't already occurred before the completion of their life cycle). The diversity of life is essentially a diversity of different life-history strategies, which life-history theory aims to explain, thus making it "*the* integrative concept of organismic biology" (Kappeler, 2021, p. 34). In the design-space of organisms, there is seemingly limitless room for the combinations of different traits that will lend themselves to different species-specific life-history strategies arising as optimal design solutions to the pathological complexity organisms are faced with.

Unfortunately, philosophers of biology with the exception of Griffiths (and his recent collaborators) have given very little attention to this theory despite it being a paradigm case of adaptationist thinking. Inspired by a talk of Griffiths in 2018 that urged us to use life-history theory as a theory of the organism as a goal-directed system to solve the problem of distinguishing healthy from pathological traits, I aim to make progress here on the goal of developing such a teleonomic theory of the organism as an agent and an account of health as a natural phenomenon by drawing on a modern extension of this theory: state-based behavioural and life-history theory.

1.4.1.1 Life-History Theory, Agency, and Adaptationism

Originally, the life-history theory was largely concerned with fairly simple models (both discrete and continuous) of the simultaneous optimization of the survival probability at different life stages and the number of offspring produced in each stage in order to maximize fitness across a lifetime (Morbeck et al., 1997; Roff, 1992, 2002; Stearns, 1992). State-dependent or state-based behavioural and life-history theory (Houston & McNamara, 1999; Mangel & Clark, 1986; McNamara & Houston, 1996) is an important extension of this theoretical framework, since it can then be used to make the notions of health and biological normativity naturalistically unproblematic by formalizing them in terms of fitness trade-offs in design. To understand an organism's teleonomic design is to understand their species-specific trade-offs between costly investments of resources into development, fecundity, and survival, with fitness providing an ultimate "common currency" for this economic decision problem, or "game" against nature. Hence, a full understanding of an organism would be an understanding of their life-history *strategy*. As John McNamara and Houston (1996) nicely put it, the life-history theory is a theory "concerned with strategic decisions over an organism's lifetime" (p. 215). It involves a naturalistically unproblematic kind of goal-directedness by treating the organisms as agents whose traits have been shaped by natural selection to contribute to a single goal of fitness-maximization:

> In life-history theory, [...] numerous aspects of an organism's life-cycle, such as the timing of reproduction or the length of its immature phase, can be understood by treating the organism as if it were an agent trying to maximize its expected number of offspring-or some other appropriate fitness measure and had devised a strategy for achieving that goal.
>
> *Samir Okasha (2018, p. 10)*

While adaptationism has often been criticized for atomistic thinking, the life-history theory is essentially a "holistic" kind of adaptationism in which organisms are not treated as a single adult phenotype, nor a mere robot-like bundle of traits (as Lewontin criticized the "adaptationists"), but a functionally complex and vulnerable life cycle – a *dynamic process* that is faced with trade-offs from birth to death, for which complex optimality problems have to be solved, in short: a life history.

It is within this theory that we can bring out the subject-side of organisms and satisfy Lewontin's demand to bring Darwinism to completion by paying attention to the "functional needs" of the organism (1985, p. 85). The external and internal factors come to be re-integrated with each other. As Pirotta et al. (2018) nicely put it, it is by "treating behavior as an evolutionary trait, [that] state-dependent life-history theory naturally integrates internal and external factors that are influencing individuals' decisions at multiple scales" (p. E52). Notwithstanding that adaptationist thinking *can* at times be misleading,[20] this book aims to show that this teleonomic theory of organismal agency provides the ideal theoretical framework to think about a major transition in organismal agency in the early Cambrian that made conscious evaluation *worth having*.

1.4.1.2 A Teleonomic Theory of Organisms

As I noted above, it is surprising that philosophers of biology have given very little attention to this theoretical framework, since it is here that we are provided with a teleonomic, or for that matter adaptationist, theory of the organism that enables us to distinguish biological normativity and functioning from all the other causal processes operating in living systems. No philosopher has made clearer the need for such thinking than Millikan (2002), who maintained that to understand life we need a way to distinguish the normative, functional, goal-directed processes within an organism from all the other causal processes, or mere "noise":

> Living chunks of matter do not come, just as such, with instructions about what are allowable conditions of operation and what is to count as allowable input. Similarly, they do not come with instructions telling which changes to count as state changes within the system and which instead as damage, breakdowns or weardowns. Nor do they come with instructions about which processes either within the organism or outside it are to count as occurring within and which are irrelevant or accidental to the system.
>
> *Ruth Garrett Millikan (2002, p. 121)*

To distinguish what matters to an organism, what is pathological, and what is part of it, one must attend to the organism's design. It is only with such a teleonomic theory of the organism that we can make sense of health and the subject-side of organisms. Physiologists may often succeed in progressing our understanding of the organism without evolutionary considerations, but once their work is intended to generalize, they ought to recognize that the various processes of the organism must be understood in "the light of life history theory" (Griffiths, 2009, p. 23).

Morbeck et al. (1997) elegantly describe life-history theory as providing us with "a means of addressing the integration of many layers of complexity of organisms and their worlds" (p. xi). This makes it the ideal agential framework to explicate the pathological complexity thesis and make sense of the evolution of subjects and their experience. In thinking about the ecological lifestyles of different species, it is thus not unreasonable to treat them as economic agents maximizing their utility (i.e. fitness). Each individual within a species is fundamentally faced with a resource allocation problem, though the solutions to this problem are admittedly more similar within a species than across species (Kappeler, 2021, p. 35). This is the economy of nature.

It is no accident that ecological and economic models share many similarities and frequently borrow from each other. Under resource scarcity, there is a constant trade-off between the parameters of reproduction and survival: "reproduction in the current age class must be traded off against reproduction later; current reproduction must be also traded off against growth, and against condition (the maintenance of structures that have already developed); current growth or condition may trade off with survival to later age classes" (Griffiths & Matthewson, 2018, p. 319). Boorse (1977) argued that one cannot build a theory of health from an evolutionary concept of adaptedness, since "parents hardly become healthier with each successive child, nor would anyone maintain that the healthiest traits are the ones that promote large families" (p. 548). But Griffiths and Matthewson (2018) are right to note that this is a very superficial take on evolution, insisting that there is a trade-off between offspring quality and offspring number as the British evolutionary biologist David Lack[21] demonstrated with his pioneering work on optimal clutch size (p. 305). It is in the context of a life-history theory that many puzzling philosophical questions about the nature of organisms and adaptation can be resolved, with fitness playing a crucial conceptual role for thinking about trade-offs in organismal design. By paying little attention to the actual work of evolutionary biologists, philosophers such as Boorse have made biologically uninformed statements about the viability of an evolutionary understanding of biological normativity.

1.4.1.3 Health and Pathological Complexity

Importantly, fitness is not just the number of offspring, though reproduction is correctly identified as the "telos" of the organism. Griffiths and Matthewson (2018) grant that Boorse's comment may have been "light-hearted" but note that "comments such as this undoubtedly contributed to the premature rejection of evolutionary views of dysfunction" (p. 305). Indeed, they gave aid to a sphere in which a serious engagement with evolutionary theory could simply be waved away with a swift remark. But fitness is ultimately the common currency for making sense of how one organism can be healthier than another. While the health of an organism is made of a vast range of different components, they do form a cohesive naturally existing state that can be expressed in terms of life-history theory. The mere fact that health is made up of multiple components is no reason to deny its existence as an organism-level phenomenon.

Only by employing the Darwinian notion of biological fitness are we provided with a common currency for weighing and combining the pathological complexity of different biological "wrongs" to assess how badly (or well) things are going for an organism. The biological world is too messy for simple binary categories, be that within consciousness or health, which is why this book naturalizes both in terms of phenomenological and pathological *complexity*. Living systems are constantly faced with trade-offs: avoidance of one danger comes at the cost of exposing oneself to another or entails foregoing some benefit. Health is simply a measure of how optimally an organism deals with the pathological complexity trade-offs it faces during its life cycle – or to put it differently, how well an organism succeeds in its species-specific life-history strategy.

Trade-offs in pathological complexity can occur on a genetic level, with selection pulling in different directions, and on a physiological level where – as Griffiths and Matthewson (2018) note – nutrients can be allocated to either somatic or germ cells. The focus here, however, will be the organism as an integrated agent. Ernst Mayr (1988) once said that "the individual and not the gene must be considered the target of selection" (p. 101). Whether this is a general truth is legitimately contested, but in thinking about the evolution of evaluative agency, it is certainly the right approach. The reason organisms do not start to reproduce from the instant of their birth onwards is a lack of immediate resources. Analogous to how a firm can be usefully treated as an agent, that must first gather resources, equipment, and manpower before selling on a market, organisms must first make an investment into growth before producing viable progeny. And these investments in, say, scales, claws, or the immune system can in turn of course have trade-offs amongst each other just like the investments of a company.

Determining how to solve the economic optimization problem between growth and reproduction under constraints was the original motivation for life-history theory. Griffiths and Matthewson (2018) point out that many organisms pursue a semelparous (as opposed to an iteroparous) strategy, i.e. "they complete all their growth before engaging in a single round of reproductive activity to which they commit all their resources" (p. 319). They mention Australian marsupials of the genus *Antechinus* as a paradigmatic example of semelparity with the death of males after a single mating season.[22] Thinking about this strange behaviour in the context of life-history theory enables us to quantify the trade-offs between survival and reproduction and think about the teleonomic design of semelparous organisms without falling prey to human-centric thinking about what makes for "good design". As I point out in an application of my pathological complexity framework, while semelparous "behaviour in males may be seen as strikingly pathological, through life-history theory we can see that it is not. Their best response to their species-specific pathological complexity is to invest all their resources into reproduction in a single breeding season, and hence this not pathological" (Veit & Browning, 2022, p. 57). Our folk understanding of health can thus be updated in the light of evolution and in particular life-history theory. Nevertheless, the *Antechinus* example is perhaps not the best illustration of an organism fine-tuned for a semelparous life history, since the mammalian body-plan is relatively unsuited to this strategy (due to the requirement for high investment into offspring, as well as an investment into an adaptive immune system which is sometimes deliberately turned off during the breeding season) but lends itself to numerous reproductive cycles.[23]

Better examples are found in insects, where many species engage in such life cycles because a semelparous strategy relies on relatively short lives, which is favoured by a high rate of juvenile survival but a high probability of death in adulthood, with bodies being discarded relatively quickly (Fritz et al., 1982). It is because of this that insects are often claimed to not feel pain, since they would not benefit from carrying such expensive equipment for their survival (see Godfrey-Smith, 2020b). This is certainly *one way* to respond to pathological complexity, though this book will argue against the notion that insects do not have hedonic valence. To a first approximation, however, it is true that many insects play a so-called r-strategy (as opposed to K-strategy), in which quantity as opposed to quality of offspring is maximized.[24] In the extreme, there is only one reproductive season before the organism dies, a strategy sometimes called "big bang reproducers" or again, semelparity as opposed to iteroparity (Diamond, 1982). Importantly, these should not be understood as binary distinctions, but rather as a continuum along a single axis with two extremes. Furthermore, whereas many mammals stop growing after reaching a reproductive stage, many insects, fish, and reptiles continue to grow until death (Griffiths & Matthewson, 2018). These are some among many other dimensions in which self-maintenance and reproduction can be in conflict. Similar life-history trade-offs can be observed in plants, fungi, and even single-celled organisms, making pathological complexity a universal design problem that emerged at the very origins of life, as a set of teleonomic problems under constraints that need to be solved simultaneously in order to maximize fitness.

1.4.1.4 Pathological and Phenomenological Complexity

As I mentioned above, life-history models were originally fairly simple – assuming that age, reproduction, and self-maintenance can be modelled independently, i.e. the particular differences between individuals at specific ages can be idealized away (McNamara & Houston, 1996; Pianka & Parker, 1975). More recent work has expanded the life-history theory to incorporate behaviour, physiology, and environmental conditions of organisms, something that is now typically referred to as state-based behavioural and life-history theory (Houston & McNamara, 1999; Mangel & Clark, 1986; McNamara & Houston, 1996). This gets us considerably closer to modelling pathological complexity as the fundamental problem of organismal trade-offs, but the problem also becomes computationally far more demanding.

To maximize fitness, it is no longer just a problem of choosing a single strategy across a lifetime, but also one of choosing strategies at any moment of one's life cycle, depending on one's bodily state and environmental conditions. The more degrees of freedom there are in the behavioural option-space (to put it simply: how many alternative actions an organism can take), the higher the pathological complexity of this fitness-maximization problem, since organisms have to make sure to make the right decisions at the right time, and this depends on their current state as well as that of the environment. Indeed, they are faced with a computational explosion of complexity. This complexity only increases when we add fluctuating environments such as changing weather conditions, food supply, risk of predation, and population density (e.g. McNamara, 1997; McNamara & Houston, 2008; Metz et al., 1992) and the frequency dependence of optimal strategies which has been extensively studied by evolutionary game theorists (Maynard Smith, 1987). These increases to pathological complexity become much harder to track, for both the organism and the biologist, but that is not to say it isn't there. Pathological complexity is the fundamental teleonomic challenge every organism has to deal with and, as I shall argue, it is precisely because of an explosion of pathological complexity during the Cambrian that sentience became a worthwhile investment, as an efficient (conscious) capacity for evaluating these continuous life-history trade-offs in action selection.

Lastly, those interested in the evolution of consciousness frequently talk of a special *lifestyle* or *mode of being* – an animal way of life emerging in the Cambrian that has given rise to minimal sentience (Ginsburg & Jablonka, 2019; Godfrey-Smith, 2020b). Using life-history theory will help us to naturalize this idea in terms of the pathological complexity of this new lifestyle and to assess the phenomenological complexity of animals we see around us here and now in terms of their distinct pathological complexity challenges. It is only within this evolutionary context of the optimal life-history strategies of biological agents that we will understand organisms as objects as well as subjects and be able to extend the Darwinian revolution towards consciousness. With this teleonomic theory of the organism in place, let me now offer an outline for the remainder of this book.

1.5 Outline of the Book

In addition to this chapter, this book consists of four further substantive argumentative chapters that each successively builds on what came before, in addition to one concluding chapter.

Chapter 2 "The Explanandum: Animal Consciousness and Phenomenological Complexity" has an important dual role. Firstly, I will show that any theory of consciousness must account for its varieties and gradations both within and across species, i.e. the explanandum of this monograph: phenomenological complexity. Secondly, I will tackle one possible objection to my project – the claim that since human consciousness is supposedly the only one we can access, it will be impossible to investigate the phenomenological complexity of animal consciousness. Drawing on a recent call by Birch et al. (2020) to develop multidimensional frameworks for animal consciousness, I will show that we already have a wide range of experimental paradigms that allow us to differentiate and measure five dimensions of phenomenological complexity, and this will allow us to move towards a comparative study of consciousness that can ultimately remove humans from the centre of reference.

By breaking the features of consciousness down into the five different dimensions of self-consciousness, evaluative experience, sensory experience, the integration of experience at a time, and the integration of experience across time, we will be able to substantially narrow the explanatory gap. It is here that Chapter 3 "The Origins of Consciousness or the War of the Five Dimensions" will further advance my goal of a bottom-up model of consciousness by trying to reverse-engineer the origin and function of consciousness – the dawn of "qualia" – by making these dimensions face off against each other to determine the most probable candidate for the first sparks of subjective experience: evaluative experience.

Having motivated my epistemic bet on hedonic evaluation as the most basic kind of subjective experience, Chapter 4 "Pathological Complexity and the Dawn of Subjectivity" will attempt to develop the beginnings of a theory by explicating the pathological complexity thesis through the resources of physiology, neuroscience, ecology, evolutionary biology, and even economics. By making sense of the lifestyle changes of animals preceding the Cambrian explosion, we will be able to explain the dawn of subjective experience in the form of a basic feel of evaluation. The evolution of animal agency required an efficient solution to the problem of handling a body with high degrees of freedom that led to an explosion in pathological complexity. This design challenge led to the evolution of hedonic valence for efficient decision-making and thus the origins of sentient beings – or, as I shall call them, "Benthamite creatures", in reference to the founder of utilitarianism Jeremy Bentham who argued not only that humans and animals pursue the maximization of their happiness or well-being but also that ethics and politics should aim at maximizing well-being, which of course has immediate implications for how other animals should be treated. As Bentham famously argued, "[t]he question is not, Can they *reason*? nor, Can they *talk*? but, Can they *suffer*?" (1879, p. 309). What motivated Bentham was the question of which organisms can subjectively experience pain and pleasure, i.e. which animals have sentience. But in doing so, he also made an important point about the best indicator for consciousness, one that unfortunately was long neglected within the denial of animal consciousness due to their alleged inability to communicate and think rationally. Bentham maintained that this cannot be what really *matters*, and it is unfortunately only in recent decades that discussions of animal consciousness have shifted away from discussions of language and intelligence towards suffering, which will be the main focus of this book.

Having constructed an alternative theory of the origins of consciousness based on a model of evaluation, Chapter 5 "Pathological Complexity Meets Phenomenological Complexity" will in turn develop phenomenological complexity as a response to pathological complexity by showing how my theory can account for the four dimensions we have previously shelved off. The pathological complexity thesis will enable us to make predictions regarding the likely subjective mental states of animals based on the evolutionary history and ecological life-history challenges faced by them *in nature*, which can then be tested through various experimental means discussed in Chapter 2. The two groups I will give the most attention to here are gastropods, as a rich source of evidence for the pathological complexity thesis, and arthropods, as a potential challenge. But I will also offer a discussion of octopuses, fish, non-avian reptiles, and corvids to show how the pathological complexity thesis can help us to scientifically approach the question of what it is like to be them.

Finally, Chapter 6 "The Final, Crowning Chapter of the Darwinian Revolution" will conclude this book by returning to Griffin's goal of moving us towards a Darwinian study of consciousness and examine how far we have come towards this goal.

Notes

1 See Avramides (2020) for an elegant review of the problem of other minds.
2 In Chapter 4, I will discuss Godfrey-Smith's thesis and its connection to mine in more detail.
3 I borrow this phrasing from an important quote of one of the 'fathers of ethology', Konrad Lorenz, that will be discussed shortly in this chapter.
4 As will become clear throughout this book, this move is inspired by similar dilemmas faced by Darwin and the early ethologists, in addition to recent work from Peter Godfrey-Smith (see Marshall & Godfrey-Smith, 2014).
5 See Browning and Birch (2022) for a recent review of animal sentience research.
6 Quote attributed to Rorty in Godfrey-Smith (2013a, p. 4).
7 Wilkes (1984), for instance, has argued so in the case of consciousness and Hesslow (1993) and Ereshefsky (2009) in the case of health.
8 See Dominic Murphy's (2020) SEP article "Concepts of Disease and Health" for an excellent recent overview.
9 Note the mirroring of the title of my dissertation "Health, Agency, and the Evolution of Consciousness" (Veit, 2023a).
10 Flanagan (1991) offers an excellent critique of these consciousness skeptics.
11 See Lewontin (1979, 1981), Gould and Lewontin (1979), and Levins and Lewontin (1985).
12 See Darwin's "Old & useless notes about the moral sense & some metaphysical points" (1840, p. 35).
13 While Spencer (1855) had already written about the evolution of mind four years prior to Darwin's *On the Origin of Species* (1859), he embraced Darwin's theory of natural selection with open arms.
14 See Godfrey-Smith (1996a) for a detailed discussion of their views.
15 See Animal Ethics (2020) for a discussion.
16 See Birch et al. (2022) for a recent special issue on this question.
17 In his biography of Wallace, Shermer (2002) offers an elegant recounting of how Wallace came to realize the phenomenon of natural selection (pp. 112–118).
18 Darwin denied that his grandfather's work had a major influence on him, but there are many similarities between their views.
19 See also Nesse (2007).
20 For critical discussions of this kind of agential thinking common to evolutionary biology, see Okasha (2018) and Veit (2021a).
21 See Lack (1947).
22 Especially the brown antechinus (*Antechinus stuartii*) (see Dobson, 2013).
23 However, as I have argued elsewhere, these animals provide an excellent case for how the pathological complexity framework can be used to evaluate not only species but also sex-dependent life-history strategies, including humans: whereas female deviations from male "norms" have often historically been labeled as "pathological" in humans, an evolutionary understanding of their different pathological complexity challenges can reveal such differences to be adaptive and thus healthy (Veit & Browning, 2022). The pathological complexity thesis may thus also help us to understand differences in subjective experiences between the sexes, but because this idea is more controversial, it will not be examined here.
24 This is not to say that the r/K framework is without problems (see Nettle & Frankenhuis, 2020).

2

THE EXPLANANDUM

Animal Consciousness and Phenomenological Complexity

2.1 Introduction

This monograph aims to make progress in our search for the place of mind in nature by considering the evolutionary origins and distribution of consciousness across the tree of life, including the question of what the subjective experiences of other animals are *like*. The goal of this chapter is to make progress on the problem of consciousness by forcing it through the theoretical bottleneck of evolutionary theory and to explicate my notion of "phenomenological complexity". Once again, let us begin with an important insight by Griffin:

> Subjective qualities and mental experiences have remained largely untouched by the Darwinian revolution, primarily for lack of effective methods for detecting them reliably in other species, let alone analyzing them by scientific methods. But, in our present state of ignorance, we certainly cannot exclude the possibility that mental experiences, like other attributes of animals and men, exhibit continuity of variation and are not typologically discrete, all-or-nothing qualities totally restricted to a single species.
>
> *Donald R. Griffin (1981, p. 125)*

If consciousness is an evolved biological phenomenon, we ought to expect both *gradations* and *varieties* of it across the tree of life – just as for any other biological phenomenon. A recognition of this phenomenological complexity will enable us to progress in our understanding of the subjective experience of other animals and narrow the explanatory gap by breaking the features of consciousness down into a number of different dimensions. This reshaping of the mind will not only enable us to resist the urge to think of consciousness as an all-or-nothing capacity that an evolutionary account is demanded to explain in one swift strike but also to acknowledge that we have already developed a wide range of experimental paradigms to investigate the minds of other animals.

2.1.1 Chapter Outline

This chapter is structured as follows. In Section 2.2 "How to Naturalize Phenomenological Complexity?", I argue that phenomenological complexity must be at the heart of an evolutionary and comparative approach to consciousness. Sections 2.3–2.6 discuss the proposed

DOI: 10.4324/9781003321729-2

multidimensional framework for animal consciousness provided by Birch et al. (2020), as well as a variety of experimental paradigms that can be used to investigate the subjectivity of other animals. This will enable us to operationalize phenomenological complexity by distinguishing five different dimensions of consciousness that can, at least in principle, all be measured in their own right. Finally, Section 2.7 "Conclusion, Objections, and Further Directions" summarizes the discussion and addresses some further difficulties in studying phenomenological complexity.

2.2 How to Naturalize Phenomenological Complexity?

In the previous chapter, I argued that a naturalization of consciousness should not rely on conceptual analysis, but instead follow Churchland's advice to bootstrap from paradigmatic cases to develop a science of consciousness. An obvious problem with this approach, however, has been the tendency to mistake the apparent features of human consciousness as insights into the necessity of all conscious experience. It is an instance of what Dennett (1991) once called the *philosopher's syndrome*: "mistaking a failure of imagination for an insight into necessity" (p. 401).

This problem is of course not unique to consciousness. As Figdor (2018) notes, "untutored imagination in general doesn't have a very good track record in terms of understanding the natural world" (p. 10). My endorsement here of an evolutionary bottom-up approach should importantly not be confused with the idea that we cannot or should not use humans as the starting point of our investigation *at all*. An insight by Lyon (2006) is important here: "Explanatory targets and starting points are not always – and perhaps not even usually – identical" (p. 52). Unfortunately, modern consciousness science has done little to distinguish human consciousness from consciousness as a natural phenomenon, suffering – as I noted in the previous chapter – from the very same problem Watson pointed out in 1913, that a focus on human consciousness would force "us into a situation similar to that which existed in biology in Darwin's time" (p. 124). However, in the face of this problem, one does not have to adopt the behaviourist response of banning consciousness entirely from science.

To study the role of consciousness in nature isn't to anthropomorphize non-human animals; it is to naturalize the mind by freeing it from the confines of the human model of consciousness (see also Figdor, 2018). Once we reject a human-centric model of subjective experience for all of consciousness, the problem of anthropomorphism will fade alongside with it. Our interest lies in what the subjective experiences of animals feel like for them, not what it would be like for us to be in an animal body. Human consciousness is a very important data point and part of what is to be explained, but it ought not to be the be-all and end-all of a Darwinian science of animal consciousness.

Evolutionary thinking is sometimes narrowly thought of as the mere construction of hypotheses about the origins and function of traits, but it is importantly also meant to be about their diversity and the wealth of alternative life-history strategies we find in nature. An evolutionary approach to subjective experience is one that recognizes both gradations and varieties of subjective experience across the tree of life, both within and across species, and studies them in a comparative manner. Consciousness, like all complex biological traits, comes in varieties and degrees, so we are bound to be misled if we use human consciousness as a *model for all subjective experience* (see also Figdor, 2018). Diversity in subjective experience across taxa is likely to be much higher than the diversity of subjective experience within a single species such as humans.

Yet, it may seem like an incredibly difficult task to study the variations and gradations of consciousness even if we limit ourselves to humans. Among humans, there are striking differences in our sensory modalities, our emotional lives, the experienced flow of time, the ability to engage in mental time travel, how unified our experiences are, and of course pathological variations of consciousness. Note, however, that care must be taken not to characterize all variations as pathological – as "lesser versions of consciousness" – since, as with any other biological trait, we should expect some variation to be the norm without thereby making it maladaptive.

Furthermore, what counts as a pathological variation of human consciousness may not be so for an animal, e.g. what would count as colour blindness in humans would not do so in a species such as the European mole (*Talpa europaea*) that evolved a less rich visual experience. Neither do we think that humans must be pathological because our distance vision is worse than that of a Peregrine falcon (*Falco peregrinus*). These examples illustrate that each species has their own biological norms and that we are in need for a teleonomic perspective that addresses the role consciousness plays for organisms in their normal lives, which was precisely what motivated Donald Griffin to call for a cognitive ethology. Understanding consciousness in all of its diversity requires a comparative study of phenomenological complexity.

However, some may maintain that such an investigation would be impossible in non-human animals, due to their inability to verbally communicate their subjective experiences. Luckily, for purposes of this book, it is precisely this challenge that has motivated Birch in a recent joint paper with two scientists at the University of Cambridge – the behavioural ecologist Alexandra Schnell and Professor of Comparative Cognition Nicola Clayton – to offer perhaps the first attempt at assessing phenomenological complexity across the animal kingdom by creating hypothetical "consciousness profiles" for other animals (see Figure 2.1).

In their article "Dimensions of Animal Consciousness", Birch et al. (2020) distinguish between what they consider to be the five most important dimensions of variation in consciousness between

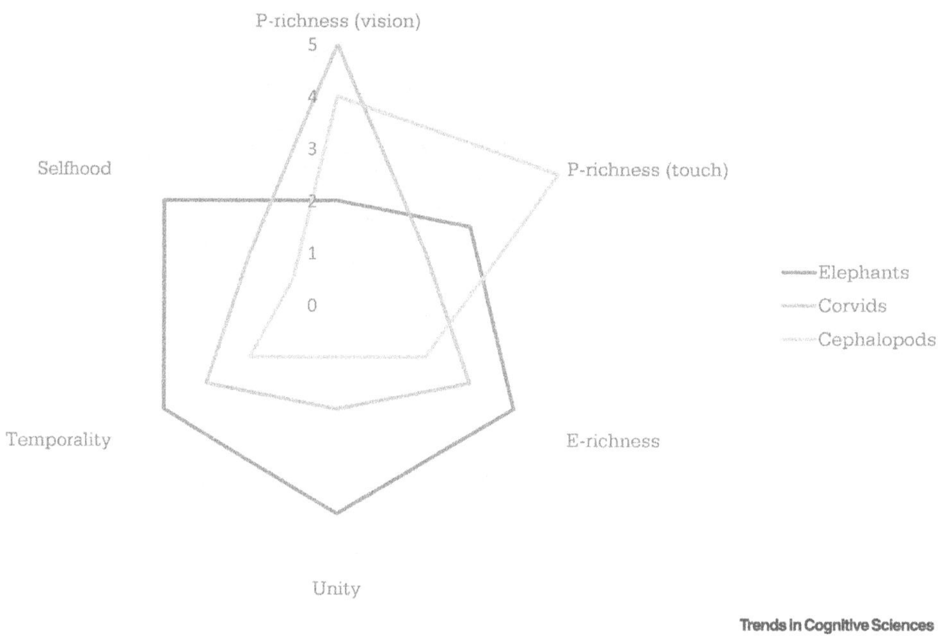

FIGURE 2.1 Birch et al.'s hypothetical consciousness profiles [reproduced from Birch et al. (2020, Figure 1, p. 791) CC BY]

"These hypothetical profiles highlight six important dimensions of variation, with p-richness represented separately for vision and touch. These are not finished, evidence-based profiles: they are conjectures based on current evidence. A key goal for animal consciousness research should be to produce a much richer evidence base for the construction of consciousness profiles and more precise ways of measuring the dimensions. Abbreviations: p-richness, perceptual richness; e-richness, evaluative richness" (Birch et al., 2020, p. 791).

species: *p-richness* (perceptual richness or rather sensory richness), *e-richness* (evaluative richness), *selfhood* (self-consciousness and an awareness of other selves), *temporality* (integration of subjective experiences across time), and *unity* (integration of subjective experiences at a specific time) (p. 790). In Table 2.1, I have borrowed their list of experimental paradigms for their proposed five dimensions, on which I will elaborate in this chapter since (i) their discussion of the dimensions is quite brief, (ii) the dimensions are important for making my case that phenomenological complexity is operationalizable, and (iii) the distinctions between the five dimensions will help us in the next chapter to reverse-engineer the evolution of consciousness and narrow the explanatory gap by re-conceptualizing consciousness as a complex multidimensional phenomenon that can gradually evolve, rather than an all-or-nothing property that would appear to resist evolutionary explanation.

2.3 The Experience of a Self

The first dimension of consciousness I will discuss here is a notion that has remained strikingly popular both in research and among the public – the idea that consciousness is self-awareness. Birch et al. (2020) call this dimension *selfhood* by which they try to capture both an "awareness of oneself as distinct from the world outside" and an "awareness of oneself as the persisting subject of a stream of experiences, distinct from other such subjects" (p. 797). Looking at different forms of selfhood in some detail will be helpful to emphasize how even within a single dimension there can be further varieties and gradations that might require further divisions.

2.3.1 Theory of Mind

We begin with what is possibly the most complex form of selfhood – the awareness of other subjects with their own distinctive subjective experience – which appears extraordinarily hard to test in animals. Yet, Birch et al. (2020) are right to point to research on the so-called theory of mind as an important line of evidence for this capacity. This "theory" or rather mindreading ability is meant to capture the capacity of mental state attribution to oneself and others (Baron-Cohen, 1997).

The prevailing paradigm for research on mindreading has largely relied on *false-belief tasks* which test whether an animal can attribute false beliefs to others (Dennett, 1987; Nichols & Stich, 2003; Saxe & Kanwisher, 2003). We know that humans have this ability, and much research has focused on the development of associated mental faculties during infancy (Dörrenberg et al., 2018; Wellman, 2014). Furthermore, the same tests have been done in those with mental disorders that appear to have mindreading deficits, such as autism (Baron-Cohen, 2000). This focus both on development and on pathological varieties is a good starting point for understanding the role of this capacity in nature.

Unfortunately, evidence for mindreading in the animal kingdom is still sparse and seen as controversial, though there is some indication that something like a theory of mind is found in other non-human primates – especially the great apes (Call & Tomasello, 2008; Krupenye & Call, 2019; Premack & Woodruff, 1978). The partial success in false-belief tasks may be indicative that many animals have a more rudimentary capacity to use their own experience to extrapolate to that of others, which Birch et al. (2020) refer to as *experience projection* (see Table 2.1).

2.3.2 Self-Awareness

When it comes to self-awareness, performance of appropriate behaviour in response to their own self-reflection in a mirror has been seen as one of the hallmark "proofs" of animal consciousness. So it is hardly surprising that Birch et al. (2020) suggest the most sophisticated current form of

TABLE 2.1 Birch et al.'s suggested experimental paradigms for the five dimensions

Dimension	*Experimental paradigm*	*Question being investigated*	*References*
P-richness	Induced blindsight	Can blindsight-like phenomena be induced in the animal through lesions to specific brain regions? If so, what information typically reaches those regions? (Drawback: highly invasive).	Cowey (2010)
	Discrimination learning	Can the animal learn to respond differently to very slight differences between stimuli (and how small can the differences be)?	Pearce et al. (2008)
	Reversal learning	When stimulus contingencies are reversed, can the animal rapidly learn that they have been reversed? This is potentially linked to consciousness in humans.	Bublitz et al. (2017); Travers et al. (2018)
	Trace conditioning	Can the animal still learn stimulus contingencies when the stimuli are separated by a temporal gap? This is potentially linked to consciousness in humans.	Clark et al. (2002); Allen (2017)
E-richness	Motivational trade-off	Does the animal weigh different needs against each other in a "common currency" to make flexible decisions?	Balasko and Cabanac (1998a, 1998b); Appel and Elwood (2009); Elwood and Appel (2009)
	Outcome devaluation and revaluation	If the value to the animal of a reward is manifestly changed, will the animal change its behaviour quickly?	Balleine and Dickinson (1998)
	Cognitive bias	Does the animal respond differently to novel stimuli depending on its affective state?	Crump et al. (2018)
	Emotional contagion	Is the animal susceptible to "catching" the emotions of other individuals?	Osvath and Sima (2014)
Unity	Interocular transfer	If the animal is conditioned to respond to a stimulus presented in one visual hemifield, can the same response be elicited by presenting it to the other hemifield?	Ortega et al. (2008)
	Meta-control	If the two visual hemifields are presented with conflicting information, can the animal resolve the conflict?	Adam and Güntürkün (2009)
	Crossmodal integration	Can the animal integrate information from different sense modalities (e.g. vision and hearing?)	Narins et al. (2005)
	Visuo-spatial bias	Does the animal exhibit visuo-spatial biases in behaviour (e.g. a preference for using a particular eye to guide a particular task?)	Schnell et al. (2016); Rogers et al. (2013)

(Continued)

TABLE 2.1 Birch et al.'s suggested experimental paradigms for the five dimensions *(Continued)*

Dimension	*Experimental paradigm*	*Question being investigated*	*References*
	Multitasking	When given two tasks simultaneously (e.g. foraging and watching for predators), does the animal divide the labour between the two hemispheres?	Rogers et al. (2013)
	Electroencephalograph studies of sleep	Does the animal exhibit unihemispheric or bihemispheric sleep?	Mascetti (2016)
Temporality (timescales <1 s)	Apparent motion	Can the animal respond differently to moving and static images? Can it make inferences from video images to real moving objects and vice versa?	Lea and Dittrich (2000)
Temporality (timescales >1 s)	Episodic-like memory	Can the animal simultaneously remember "what", "where", and "when" about a specific past event?	Clayton and Dickinson (1998)
	Source memory	Can the animal remember information about how a memory was acquired (e.g. by vision or by smell)?	Billard et al. (2020)
	Memory integration	Can the animal update old memories with new information?	Clayton et al. (2001)
	Future planning	Can the animal flexibly and spontaneously plan for a future event, and for future desires, without relying on reinforcement learning?	Cheke and Clayton (2012)
Selfhood	Mirror-mark	Does the animal recognize a mark seen in a mirror as a mark on its own body?	Anderson and Gallup (2015); Morrison and Reiss (2018); Plotnik et al. (2006); Prior et al. (2008); Kohda et al. (2019)
	Body awareness	Can the animal recognize the position of its own body as a potential obstacle to success in a task?	Dale and Plotnik (2017)
	Experience projection	Can the animal predict how others are likely to behave in a scenario on the basis of a specific past experience it had in the same scenario?	Emery and Clayton (2001); Kano et al. (2019)

Source: Reproduced from Birch et al. (2020, Table 1, p. 798) CC BY.

"A list of established experimental paradigms with the potential to provide insight into p-richness, e-richness, unity, temporality, and selfhood. There is continuing debate regarding the implications of these paradigms for questions about conscious experience. Inferences to properties of conscious states will be stronger when based on a battery of convergent experimental results from different paradigms. We restrict our attention here to established paradigms [...]" (Birch et al., 2020, p. 798).

this test – the *mirror-mark test* – as the key to studying this capacity (see Table 2.1). Here, an animal is tested in regard to whether it "is able to recognise a mark seen in a mirror as a mark on its own body" (Birch et al., 2020, p. 797), i.e. whether they are able to make a mental connection between the "self in the mirror" and their own body as a self in the world.

Most human infants by around two years old are able to pass the test by touching the mark as guided by their reflection in the mirror, though there is some inconsistency in the results due to varying operationalized definitions (Archer, 1992; Bard et al., 2006). But this should be expected in a gradualist picture; already in healthy human development we recognize that this capacity doesn't just pop into existence, so we shouldn't be surprised if this side of experience is reported to vary across the animal kingdom. Birch et al. (2020) even highlight a recent study by Kohda et al. (2019) on the cleaner wrasse (*Labroides dimidiatus*) to suggest that "grade of self-consciousness required to pass the mirror-mark test is possessed by a wide range of animals" (p. 797).

What Kohda et al. (2019) demonstrated was that the cleaner wrasse not only behaved atypically in front of the mirror, or ceased social behaviour towards the reflection, but also attempted to remove a coloured mark through scraping behaviour, which they did not do when the mark was translucent or in the absence of a mirror. This suggests a higher level of self-awareness than is typically assumed in fish. Naturally, there has been much controversy surrounding their experiment and the mirror-test paradigm at large (Gallup Jr & Anderson, 2020; Vonk, 2019) which relates to the parallel controversy over the "correct" meaning of self-awareness (see Gallagher, 2011 for a number of different definitions).[1]

Much of the controversy here results from a mistaken binary conception of self-consciousness (which is partially encouraged by the very idea of the mirror *test* that can be passed or failed, as introduced by Gallup, 1970). Self-consciousness, like other mental phenomena, does not just suddenly pop into existence like a light being switched on, but gradually emerges in incremental steps towards greater cognitive and phenomenological complexity (de Waal, 2019; de Waal & Ferrari, 2010; Godfrey-Smith, 2020c). As Frans de Waal (2019) describes it, we still "live with a [']Big Bang['] theory, according to which this trait appeared out of the blue in just a handful of species, whereas the vast majority lacks it" (p. 1).

Borrowing de Waal's mirror self-recognition (MSR) diagram to distinguish the two views, it is obvious that what he calls the "traditional binary model" (**A** in Figure 2.2) fails to recognize gradations of self-awareness. Here, self-awareness is only attributed to the great apes, elephants, dolphins, and magpies, who are known for their high intelligence and problem-solving ability, whereas *all* others are lumped together into a single category considered to lack self-consciousness. This gross simplification of the empirical data results in much information being lost. In contrast, a gradualist view of self-consciousness recognizes a wide variety of self-recognition abilities across taxa (**B** in Figure 2.2).

All hominids appear, de Waal (2019) points out, to "spontaneously explore and play with their reflection and care about their appearance" (p. 5). Furthermore, we can recognize a broad range of intermediate levels of response, where we could place African grey parrots (*Psittacus erithacus*) and monkeys (*Macaca tonkeana* and *M. fascicularis*) that have been shown to use a mirror as a tool to their advantage in order to discover objects outside of their visual field (Anderson, 1986; Pepperberg et al., 1995). Even with this decades-old paradigm, we are already able to recognize a diversity of more and less sophisticated forms of self-awareness:

> Reactions to mirrors range from permanent confusion about one's reflection to a certain level of understanding of how mirrors operate (e.g., using them as tools) and only brief or no confusion between one's reflection and a stranger.
>
> *Frans de Waal (2019, p. 5)*

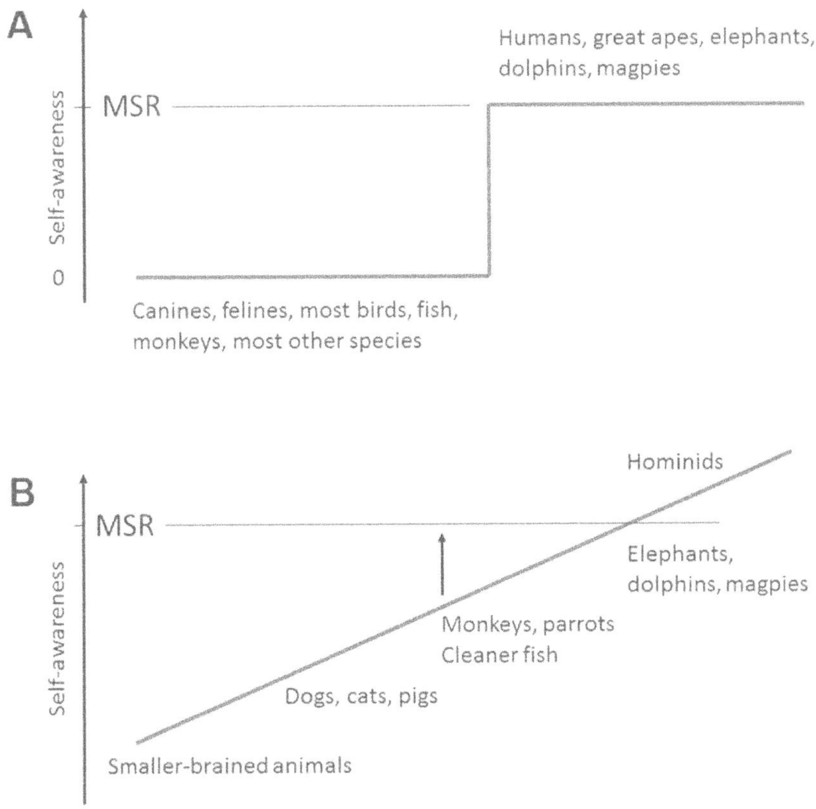

FIGURE 2.2 A binary (**A**) versus a gradualist (**B**) view on the evolution of self-awareness [reproduced from de Waal (2019, Figure 3, p. 5) CC BY]

Moreover, de Waal (2019) urges us to recognize that multiple sensory modalities are involved in such findings, since marks often involve a physical object, e.g. a sticker, rather than just a discolouration and are thus visually much more noticeable and abnormal, in addition to plausibly causing pain and discomfort, or at least some somatosensory experience.[2] This is precisely why de Waal included an arrow in his depiction of the gradualist view of self-awareness (**B** in Figure 2.2), since training and multimodal stimulation are able to raise an animal towards displaying a higher level of self-awareness. A visual mark may simply not be interesting to many animals whose lifestyle naturally leads to discolourations.

For cleaner wrasse, however, visual marks are likely to be ecologically salient; the species frequently engages in cooperative social behaviour with larger host fishes, having evolved and become adapted to detecting and freeing these host fishes from parasites and dead skin. This then makes it less surprising that they are able to pass the visual mirror-mark test. Such a social lifestyle, as de Waal (2019) notes, involves complex "economic decision-making" (p. 2). Hence, within the context of my framework, an increase in phenomenological complexity in the face of this pathological complexity should not at all be surprising.

These varieties and gradations of self-awareness elegantly illustrate the importance of locating empirical findings in tests about consciousness within an evolutionary context. As de Waal (2019) puts it: "Only with a richer theory of the self and a larger test battery will we be able to determine all of the various levels of self-awareness, including where exactly fish fit in" (p. 7). To assume that a single experiment – or for that matter a set of experiments – could simply be applied across the tree of life without any consideration for the species' ecological lives and evolutionary history is as absurd as using the mirror-mark test on a pig or hippopotamus (Allen, 2004, p. 598). It is here that the pathological complexity thesis will offer us a useful framework for accommodating the criticism commonly raised by cognitive ethologists that the animal's ecological life-history context is being ignored within comparative psychology.

2.3.3 A Minimal Sense of a Bodily Self

Lastly, the most minimal form of self-consciousness Birch et al. (2020) plausibly suggest involves "registering a difference between self and other: registering some experiences as representing internal bodily events and other experiences as representing events in an external world" (p. 797). We may want to call it a bodily self. Such a capacity they suggest (similarly to de Waal) is plausibly found in all complex animals that need to be able to distinguish their own sources of bodily feedback from external ones. In addition, we may want to include features such as an awareness of oneself in space and perhaps an awareness of one's bodily state that in richer forms eventually give rise to self-awareness. Birch et al. (2020) mention *body awareness* experiments, such as those by Dale and Plotnik (2017) on elephants that try to test whether an animal is able to recognize its own body as an obstacle in a task (see Table 2.1).

Yet, determining how to distinguish conscious self-recognition from unconscious self-recognition would require a better theoretical understanding of the function of consciousness, which makes it hard to have agreement on what kinds of tests would vindicate a minimal feeling of bodily self – a problem that is admittedly shared with all the minimal forms of the five dimensions. Since we are interested in the origins of consciousness, the next chapter will look again at this minimal form of selfhood and its possible connection to consciousness, but we shall now turn to the dimensions of unity.

2.4 The Unity of Experience

Birch et al. (2020) understand unity as the integration of experience *at a point in time*. By this, they mean the experience of a single self or subject as opposed to multiple selves. The terminology is not perfect, however, since they use the label *temporality* for the unification of experience across time. Here, I will simply discuss them as two dimensions of unity: one I shall call synchronic and the other diachronic.

2.4.1 Synchronic Unity

We know from our own experience that healthy human adults have a highly unified experience. Smell, taste, vision, general bodily feelings – whatever it is we're experiencing appears to be experienced synchronously. Other animals, however, may not have such a unified experience.

As Birch et al. (2020) point out, the curious case of split-brain patients already suggests that the unity of consciousness may not be a necessary feature of subjective experience (see also Godfrey-Smith, 2021a). In the twentieth century, some people suffering from severe forms of epilepsy had their *corpus callosum* (the connecting pathway between the two brain hemispheres) cut as an experimental treatment (Schechter, 2018). Those who had it "wholly or partially severed", however, occasionally displayed strikingly "disunified behaviour when different stimuli are presented to the two halves of the visual field" (Birch et al., 2020, p. 793). Indeed, such split-brain patients appear to have two distinct streams of experience:

> If these subjects are asked to verbally describe what they see, they will report what is visible on the right-hand side of their visual field. This is because language is predominantly controlled by the brain's left hemisphere, which only has access to visual information from the right-hand side. Yet, when asked to draw with the left hand what they see, they will draw what is visible on the left-hand side of the visual field. This is because the left hand is predominantly controlled by the right hemisphere, which only has access to visual information from the left-hand side.
>
> *Birch et al. (2020, p. 793)*

Researching decision-making in split-brain patients can help us to understand how a healthy brain achieves what Birch et al. (2020) call *meta-control* when faced with conflicting sensory information from the hemispheres (see Table 2.1), which may hold clues regarding the role of unity. In animals with highly lateralized brain hemispheres, however, disunified experience will not necessarily be dysfunctional, so the study of this capacity in animals whose brains resemble those of split-brain patients may reveal the adaptive benefits of synchronic disunity.

As Birch et al. (2020) point out, for animals that simultaneously engage in multiple complex tasks, i.e. what they call *multitasking* (see Table 2.1), there might be a division of labour between hemispheres. A useful experimental paradigm Birch et al. (2020) mention that could probe such division of labour are *visuo-spatial bias tests* (see Table 2.1) that check whether there are preferences for using a particular sensory organ to deal with a task. Relatedly, Birch et al. (2020) suggest that *crossmodal integration* (see Table 2.1) could help to assess whether information is being integrated differently depending on the sensory modality (e.g. touch vs pain). After all, unity may not be an all-or-nothing affair.

What I consider to be the most important paradigm that Birch et al. (2020) discuss for this dimension is the study of *interocular transfer* (see Table 2.1). Here, an animal is trained "to perform a task in response to a stimulus presented to one eye", after which we test "whether the task can still be performed when the stimulus is presented to the other eye" (Birch et al. 2020, p. 793). If an animal fails to do so, this could be suggestive of disunified experience, where the different brain hemispheres contain in some sense distinct "selves".

Finally, another compelling paradigm mentioned by Birch et al. (2020) for studying the unity of conscious experience is *electroencephalograph studies of sleep* (see Table 2.1). By studying the electrical activity of brains during sleep, we can determine whether one of the hemispheres is sleeping while the other is in a wakeful mode, i.e. whether they have bihemispheric or unihemispheric sleep. As Birch et al. (2020) rightly note, this evidence would be suggestive of "more than one stream of consciousness" (p. 794). We will return to this and other experimental paradigms in the next chapters to discuss the connection between synchronic integration and consciousness. Let us now consider the dimension of temporal unity.

2.4.2 Diachronic Unity

Birch et al. (2020) call the integration of experience across time *temporality* and illustrate it using the example of subjectively experiencing "the leaves of a tree blowing in the wind", which we do not experience as a mere "series of static snapshots" (p. 794). Beyond this integration of experience across short timescales, however, Birch et al. (2020) also note that humans are able to engage in *mental time travel*, i.e. the ability "to recall past experiences and simulate future experiences" (p. 794). In their table of established experimental paradigms, Birch et al. divide those paradigms testing timescales of under one second against those that integrate over a longer time span (Table 2.1). Naturally, the second category is a much more expansive one and appears to be related to memory, whereas the first is about the way we experience time. Let us thus divide them into mental time travel and the experience of time.

2.4.2.1 Mental Time Travel

Most animals are assumed to lack the ability for mental time travel – indeed, they might well be described as "living in the moment". So it is hardly surprising that the study of mental time travel in animals is a controversial subject and Birch et al. (2020) suspect that the ability would require "substantial cognitive sophistication" (p. 794). However, as they point out, we have already accumulated a wealth of data suggesting that corvids have this capacity, which I will discuss in Chapter 5.

There are plenty of tests for the backwards-looking aspect of conscious mental time travel or as it is typically called "episodic memory". Birch et al. (2020) mention a paradigm that Clayton herself was involved in developing, i.e. testing for *episodic-like memory* (see Table 2.1), which is meant to assess whether animals are able to retrieve memory about the so-called three Ws: "what", "when", and "where" of past events. However, the suffix "-like" was added precisely because success in such tasks does not necessarily imply conscious recall.

Birch et al. (2020) recommend that tests of *source memory* (see Table 2.1) may help us to demarcate information that can be retrieved consciously, by investigating whether the animal can recall the sensory modality through which the information was acquired, rather than just the content. For this, the suggestion by Birch et al. (2020) to study *memory integration* (see Table 2.1) is also helpful since the ability to update old memories in light of new experiences could be suggestive of conscious mental time travel.

Lastly, in regard to mental time travel directed towards the future, Birch et al. (2020) suggest research on *future planning* as flexible future-directed behaviour that goes beyond mere reinforcement learning (see Table 2.1). If we can find such planning, Birch et al. (2020) are certainly right to consider it as a "promising nonverbal indicator of conscious temporal integration" (p. 796).

2.4.2.2 The Experience of Time

Could some animals perceive the world in a more fragmented way than humans? This question may seem impossible to answer. Yet, Birch et al. (2020) suggest that we could draw on research on the so-called colour-phi illusion, "in which two spatially separated, differently coloured dots flashed in sequence are perceived as a single moving dot that changes colour half-way across the gap", to understand how the brain integrates them to construct "a coherent account of how the stimulus is changing" (p. 794). This research supports the idea that integration of sensory stimuli plays an important functional role in humans. In some animals, however, things might look different.

Birch et al. (2020) urge us to take seriously the possibility of conducting colour-phi tests in non-human animals, even in the absence of a verbal report, by suggesting that we could train animals to "respond differently to perceptions of continuous and discrete stimuli and to stimuli that change colour half-way and stimuli that do not" (p. 794). If they were then to be presented with a colour-phi test, we could change the interval between both stimuli in order to test whether – just like in the human case – there is a threshold at which animals move between an experience of the phenomena as two discrete or one continuous stimulus. Similarly, we could draw on what Birch et al. (2020) call the *apparent motion* paradigm (see Table 2.1) to assess how animals react to videos with higher and lower frame rates.

An experimental test that they surprisingly do not mention – despite already having been investigated extensively in animals and likely constituting the best evidence for assessing the experience of time in other animals – is the *critical flicker fusion frequency* paradigm, which tries to find the lowest frequency of a flickering light at which the light is perceived as continuous rather than as independent stimuli (D'Eath, 1998; Healy et al., 2013; Potier et al., 2020).[3] A broad comparative study of this capacity may reveal the evolutionary benefits and drawbacks of a more fragmented way of experiencing the world, the possibility of which I will discuss in the next chapter, but we shall now turn to the dimension most readily associated with consciousness research.

2.5 Sensory Experience

The perhaps most diverse dimension of phenomenological complexity is sensory experience. Birch et al. (2020) describe the dimension of sensory experience as *perceptual richness* (or p-richness). I prefer the term sensory richness because it is less evocative of a primacy of the modality of vision, which has unfortunately so far dominated much of consciousness science. Following Birch et al. (2020), we can understand different degrees of richness in terms of how fine-grained the conscious (sensory) discriminations of an animal are.

This doesn't mean, however, that all sensory modalities can be put on a single scale. Unlike with the other dimensions, Birch et al. (2020) maintain that there are as many sub-dimensions of the sensory side of consciousness as there are different sensory modalities, such as vision, smell, and touch. Furthermore, there are sensory modalities that some animals have but humans lack (Keeley, 2002). So it is perhaps unsurprising that Birch et al. (2020) deny the possibility of an overall measure for how different sensory modalities could be integrated into an overall measure: "Any measure of p-richness is specific to a sense modality, so we should not refer to a species' overall level of p-richness. A species might have richer perceptual experiences than another in one modality, but less rich experiences in a different modality" (p. 790). This is nicely illustrated in Figure 2.1 which distinguishes the hypothetical consciousness profiles of elephants, corvids, and cephalopods by splitting p-richness into vision and touch. To their credit, Birch et al. also accept that perceptual consciousness can vary within a sense modality, which (though they don't state this explicitly) could then lead to a sentience profile for a particular perceptual dimension of an animal.

Visual experience, Birch et al. (2020) note, could be divided into "bandwidth (the amount of visual content experienced at any given time), acuity (the number of just-noticeable differences to which the animal is sensitive), and categorisation power (the animal's capacity to sort perceptual properties into high-level categories)" (pp. 790–791). But their reply to the worry that this would make an overall evaluation difficult is not particularly satisfying, since they merely maintain that an animal can be considered to have a richer visual experience if it outperforms another in two but loses in one. But surely this would also depend on the case: consider a species that is mildly worse in two categories but outperforms another species in the third by a large margin. Nevertheless, Birch et al. accept that further empirical investigation may require us to move to an even more fine-grained profile or – as I would put it – a realization that there is greater phenomenological complexity in nature than we may have anticipated.

The most basic paradigm that Birch et al. (2020) mention for measuring p-richness is *discrimination learning* (see Table 2.1), which tests whether animals are able to respond to ever more fine-grained differences between different stimuli. The cognitive capacity to distinguish stimuli alone, however, may not provide us with evidence that they are also consciously experienced. To address this problem, Birch et al. (2020) suggest relying on blindsight research that has received much discussion in the consciousness literature, since those suffering from this pathological condition "report blindness in part of their visual field, but they are able to use visual information about objects in that region to guide action" (p. 791).[4]

Birch et al. (2020) argue that we could use experimentally *induced blindsight* (see Table 2.1), which has been commonly used to target the primary visual cortex (V1) of monkeys, who appear to behave strikingly similarly to humans with blindsight in *forced-choice tasks* when this part of their brain is damaged. These findings, Birch et al. (2020) argue, could be compelling evidence for sensory consciousness in animals: "if a stimulus is processed in a brain region such that damage to that region results in blindsight, then a healthy, blindsight-free animal of the species in question probably perceives that stimulus consciously" (p. 792). On pains of evolutionary continuity, this method appears quite promising. Indeed, Birch et al. suggest that we could extend this strategy "to non-mammals, based on identifying homologues or analogues of V1 in those animals" (p. 792).

Yet, they are right to admit that experimentally induced blindsight is highly invasive, difficult, and understandably considered ethically problematic despite its scientific potential. However, not all neuroscientific methods of inducing blindsight must be invasive, as recent studies on *transcranial magnetic stimulation* of the visual cortex have shown (Christensen et al., 2008; Jolij & Lamme, 2005; Ro et al., 2004). While Birch et al. surprisingly do not mention this method, it may provide an excellent way to temporarily induce blindsight-like states by using magnetic impulses to disrupt the local brain regions responsible for visual experience in other animals (see Railo & Hurme, 2021 for a critical review).

Another alternative method that Birch et al. propose for making progress on assessing the sensory worlds of animals is to focus on identifying special tasks that have been linked to conscious experience in humans and apply corresponding tasks to animals. One of their proposed paradigms for this is *reversal learning* where the valence of the reward associated with two different stimuli is switched and a measure taken of the time until an animal learns the new connection, which appears to have links to conscious perception in humans (see Table 2.1).

The experimental paradigm that Birch et al. highlight for a cognitive test of sensory consciousness is *trace conditioning*, where a conditioned stimulus and an unconditioned stimulus are presented with a short delay between them to check whether the connection can be applied over these timescales as opposed to merely instantaneously (see Table 2.1). As they point out, in trace conditioning tests on humans, subjects appear to only produce a conditioned response if they have been consciously aware of the initial unconditioned stimulus (e.g. a sound), which suggests that this form of learning is indicative of consciousness and could be applied in non-human animals. Since tests involving learning are also related to evaluation, let us now turn to the last of the five dimensions.

2.6 Evaluative Experience

As I identify the origins and function of consciousness within hedonic valence, it is unsurprising that the dimension of evaluative richness (e-richness) plays a special role in this book and will hence receive extra attention here. While philosophers have given relatively little attention to this dimension, several well-known scientists continue to defend the intuition of Romanes that an understanding of the evaluative mechanisms of the brain would bring us closer to understanding consciousness and suffer less from the challenge coming from the hard problem. Most prominently, the father of affective neuroscience (the neuroscience of emotions) Jaak Panksepp (1998, 2005, 2010, 2011) argued that emotions plausibly constitute the most basic and ancient kind of consciousness – a view that will also be defended in the following chapters.

For Birch et al. (2020), this evaluative dimension of consciousness is meant to capture the subjective experience of emotions and moods – which are often described as so-called affects (Browning, 2020b; Panksepp, 2005). They suggest that "valence" would be a great concept for understanding varieties in animal experience, since it is always involved in affective (emotional) decision-making, irrespective of speculations about whether the animal exhibits complex human-like emotions. I am less convinced of this statement since the terms "valence" and "affective decision-making" are used in an ambiguous way in the literature in both conscious and unconscious senses of evaluation. What they appear to have in mind are rather hedonic decision-making and hedonic valence:

> Some conscious emotions, such as pain, fear, grief, and anxiety, feel bad. These are affective experiences with negative valence. Others, such as pleasure, joy, comfort, and love, feel good. These are affective experiences with positive valence. All affective responses have positive or negative valence.
>
> *Birch et al. (2020, p. 792)*

Importantly, my emphasis on adding the prefix "hedonic" is not a mere side note. Research in animal welfare science and the affective sciences (the sciences of emotion) has notoriously suffered from ambiguity in the use of terms such as "fear", "valence", "emotion", and "welfare", where these are sometimes used as descriptions of nonconscious neurological, behavioural, or physiological processes and at other times used to refer to subjective experiences. Prominently, this ambiguity has recently been criticized by Dawkins (2017a, 2017b, 2021) in animal welfare science and by LeDoux (2017a, 2017b, 2019, 2022) in affective neuroscience, both of whom think that this confusion has harmed the progress of their sciences and urge caution when talking about consciousness in non-human animals.

Dawkins (2017a) has described this ambiguous use of language as "flirting with consciousness", i.e. researchers use terms that in their everyday use carry implications of subjective experience, but when pressed in a scientific context tend to back off and state that they are not necessarily implying consciousness. Because of this, LeDoux and Dawkins believe that both the public and scientists have come to overestimate how much we really understand about conscious emotions, especially in other animals. Notably, LeDoux (2022) himself admits that he has been guilty of this way of speaking in his earlier pioneering work on how the brain achieves Pavlovian fear conditioning in rats, which has led to him routinely being introduced as the discoverer of "how conscious feelings of fear arise from the amygdala" (p. 4). LeDoux suspects that the origins of this ambiguous language lie in the behaviourists' usage of folk mental state terms (e.g. "fear" conditioning) for what they saw as purely behavioural learning processes, not conscious emotions. But because the behaviourists failed to communicate their definition to the public, and more and more anti-behaviourists, such as Panksepp (2005, 2011), were interested in actually studying fear in its folk sense as a conscious experience, LeDoux (2022) asserts that we have returned to the "semantic 'wild west' of the late nineteenth century" when "animal psychology tried to use intelligent and emotional behaviour as marks of consciousness" (p. 8). As a solution, LeDoux (2022) has proposed to use terms such as "fear" exclusively for the subjective experience it refers to, rather than its "behavioural and physiological correlates" (p. 4). What he once called "fear circuits", he thus now calls "survival circuits" (LeDoux, 2022).

At a first glance, the use of mental-state neutral language for what goes on in animals may appear sensible. However, in doing so, LeDoux ends up artificially widening the explanatory gap between the mechanisms of the brain and the "mysterious" subjective feelings of the mind. Admittedly, LeDoux and Dawkins are right to warn that it is very difficult to establish links between brain mechanisms and subjective experience and that, because the brain does a lot unconsciously, we are in need of an explanation as to why a particular brain process is conscious rather than nonconscious. However, many, if not most, animal consciousness researchers are well aware that most information processing in the brain is done unconsciously and that we have to disassociate conscious from unconscious processes. The fact that researchers such as Panksepp made *hypotheses* identifying subjective experiences with particular brain processes should not at all be seen as a "Gotcha!" moment revealing the naiveté of the field. The science of consciousness has made progress thus far precisely by challenging proposed hypotheses for the functions of consciousness and by showing that many proposed functions can also be performed unconsciously.

This "speculation" is simply how science progresses, not a deep conceptual and methodological error. In a biological materialist picture of consciousness, subjective experience is constituted by particular mechanisms and processes of the nervous system. It is not "produced" by it – such thinking would lead us straight back to dualism. But the insistence that "flirting" with consciousness must be a bad thing and that we should not use any mental state terms for physical processes inevitably brings with it older dualist ways of thinking. Similarly, the assertion by LeDoux (2019) that the problem of other minds is merely "a hypothetical philosophical argument, not a scientifically based one", and only applies to non-human animals as they have very different brains from our own, likewise evokes strangely dualist thinking about the separation of philosophy and science and the difference between us and other animals.

LeDoux (2019) is, of course, entirely correct in stating that "if the unique aspects of our brain and our cognition are key to our kind of consciousness, then our kind of consciousness should not simply be assumed in other animals on the basis of the other minds problem" (p. 318). But his implication that the work of contemporary animal consciousness researchers is no different from nineteenth century speculations that simply anthropomorphized other animals and attributed human-like conscious experiences to them based on some behavioural similarities is an unfair straw-manning and misrepresentation of contemporary animal minds research:

> When the claims match common sense and lore, they feel correct, and when they are repeated authoritatively in scientific or lay communities, they come to be assumed as indisputable facts. [...] The widespread assumption that innate defensive behaviours are a fool-proof reflection of conscious feelings of fear is a case in point.
>
> *Joseph LeDoux (2022, p. 8)*

These assertions are greatly overstated. While members of the public might anthropomorphize their own pets, few would accept innate defensive behaviours in insects, gastropods, and worms as foolproof evidence for consciousness. Even for more complex evaluations and survival behaviour, few animal consciousness researchers would accept these as foolproof indicators of conscious feelings. Indeed, there is hardly a scientific field that is kept under higher scrutiny for its hypotheses about its target phenomenon. Yet, LeDoux (2019) simply asserts that animal consciousness researchers fail to live up to the standards of human consciousness research, since their "experiments are not designed to ascertain whether a particular behaviour is consciously or non-consciously controlled. Instead, they involve amassing more and more support for the intuition that consciousness was involved" (p. 319). Worse, he accuses animal consciousness researchers of (i) not knowing the methods for distinguishing conscious from unconscious processes, (ii) not caring about them because they would support nonconscious explanations, and (iii) endorsing animal consciousness for moral rather than scientific reasons (LeDoux, 2019, p. 320). I suspect that many contemporary animal consciousness researchers would take great offence at these accusations.

As this chapter hoped to illustrate, animal consciousness researchers care a great deal about disassociating conscious mental processes from unconscious ones, possessing a rich diversity of empirical approaches that put them in a very different position from the speculations of Darwin, Romanes, and others. Although future research will inevitably improve on the methods I am discussing here, they are not just the non-verbal guesswork LeDoux tries to make them out to be. That the origins of animal consciousness research have involved plenty of speculation should be seen as no more problematic than the speculations we find at the origins of any new scientific field of research, so long as they are empirically grounded and provide us with eventual tests with which to examine hypotheses.

Nevertheless, one might readily see the recent book by LeDoux (2019) as an inverse project of the one I am engaged in here. Whereas I argue that consciousness arose very early in the history of animal life in order to deal with the fundamental trade-off problem of life (i.e. pathological complexity), LeDoux spends a large part of his recent book on the deep history of survival behaviours, from single-cell organisms to us, to highlight that we cannot simply infer consciousness from the presence of such behaviour and that consciousness should therefore be seen as a much more recent invention. While the survival circuits and survival behaviours of animals have become more complex over evolutionary time, he thinks that this is no reason to attribute sentience to them, since they are mere "manifestations of an ancient survival function – the ability to detect danger and respond to it" (p. 1). Here, LeDoux (2019) simply commits the same mistake as the biopsychists who use the evaluative behaviour of single-cell organisms to argue that all life is sentient. As Chapter 4 will make clear, both fail to think in evolutionary terms and recognize important gradations and major transitions through which consciousness could have gradually evolved. The presence of evaluation in nonconscious organisms cannot be used as a "proof" that *all forms* of evaluation must be nonconscious. The goal of the pathological complexity thesis is simply to defend one such hypothesis about a special kind of teleonomic complexity that made a hedonically felt form of evaluation worth having and led to a major transition in the evaluative agency. While LeDoux and Dawkins may think that any such hypothesis for animals could also be explained through nonconscious means, I will argue in Chapter 4 that hedonically felt valence is the best explanation of how animals came to deal with just this kind of complexity and thus how consciousness gradually evolved.

But before I do so, let us first look at some of the empirical methods that have been developed for assessing "valence" and "affective states" in other animals. This is important, whether or not they are consciously experienced, since it is only by developing a greater understanding of the evaluative capacities of animals that we will be able to make sense of the evolution and function of sentience. As Birch et al. (2020) nicely put it: "Finding out how positive and negative valence are produced in an animal, and how these processes vary across taxa, should be a central goal of animal consciousness research" (p. 792).

A particularly useful paradigm Birch et al. (2020) mention to investigate the role of affects is work on *cognitive bias* in animals, i.e. whether an animal reacts differently to new stimuli based on which affective state they are in (see Table 2.1). Relatedly, their proposal to study *outcome devaluation and revaluation* (see Table 2.1) investigates whether and how quickly an animal will change its behaviour when the rewards associated with a behaviour are changed, which is also related to the reversal learning discussed in the sensory dimension. Finding evidence for these capacities could be suggestive of evaluative experience.

Furthermore, they also list *emotional contagion*, i.e. the "transmission" of affective states across individuals, as a promising paradigm for evaluative richness (see Table 2.1). Among highly social species that require synchronous behaviour, it may even be possible to find richer evidence for this capacity than is found in humans, but this paradigm has unfortunately been understudied.

Naturally, here we find a close connection to other minds research discussed in the dimension of self-consciousness.

The most significant experimental paradigm for evaluative experience, however, that Birch et al. (2020) introduce is that of *motivational trade-offs* (see Table 2.1). It is the study of how animals weigh multiple needs and opportunities against each other. Unfortunately, there has been very little comparative work on this capacity, but Birch et al. (2020) declare this paradigm a "priority for future work" (p. 793). Indeed, the importance of trade-offs is one of the core motivations for the pathological complexity thesis and will be looked at in detail in Chapter 4, but I shall lay some of the groundwork here.

Importantly, Birch et al. (2020) argue that valence provides an evaluative *common currency* for just these kinds of trade-offs in affective decision-making (p. 792). This is a different usage of the term "common currency" from the way I have used it in Chapter 1, but they are conceptually very similar. Unlike fitness, which provides a common currency for the trade-offs of pathological complexity that organisms face in their life histories, this usage of a common currency refers to a common currency that helps organisms to evaluate trade-offs arising from their different needs/goals. There are thus two domains where common currency arguments are frequently made, one in evolutionary biology, where fitness provides an *ultimate* common currency for optimal design, and the other in the behavioural and cognitive sciences, where people talk of a *mechanistic* common currency for optimal decision-making. Perhaps unsurprisingly, many authors in the behavioural, cognitive, and affective sciences have argued that complex animals have, or perhaps even must have, a *proximate* common currency linked to fitness in which the values of different actions are ranked (Cabanac, 1992; McCleery, 1977; McFarland & Sibly, 1975; McNamara & Houston, 1986; Shizgal & Conover, 1996).

Again, however, we should point out that this mechanistic claim is defended in both conscious and nonconscious versions and Spurrett (2014) has offered an elegant paper criticizing the common conflations of these common currency claims both in science and in philosophy. Indeed, the idea that hedonic valence constitutes a felt common currency of decision-making is an old one that has long been defended by philosophers in the tradition of psychological hedonism and utilitarianism, such as Bentham (1879). Here, it is also important to distinguish between the usage of pleasure as a particular kind of mental state with positive valence (as done in the above quote of Birch et al.) from pleasure as positive valence itself. Utilitarianism has often been attacked for narrowing all of the human experience down into pleasure and pain, but Bentham was using them in the deliberately broad sense of hedonic valence common to all affective experience: the pleasure*s* and pain*s* of experience. This is the sense in which Chapter 4 will explicate the evolution of Benthamite creatures.

Furthermore, the idea that all affective states in humans and non-human animals can be mapped onto a two-dimensional space of valence and arousal is a widely shared view with a long tradition in the study of emotion, in both psychology and animal welfare science (Burgdorf & Panksepp, 2006; Mendl et al., 2010; Plutchik, 1962, 1980; Russell, 2003; Russell & Barrett, 1999; Russell & Fehr, 1987). Rather than thinking of emotions in terms of "typological" theories that treat them as discrete entities, such as in the work of Panksepp (2005), this tradition of "dimensionalist" approaches treats the dimensions of valence and arousal as the real phenomena in nature, with emotions being mere theoretical constructs aimed to target particular points in these dimensional spaces.[5] Today, these ideas are especially important in Lisa Feldman Barrett's (2017) "constructivist" theory of emotions, which had a major impact on affective psychology, in both humans and non-humans. Naturally, my approach bears greater resemblance to the tradition of the dimensionalists, rather than the typologists, since we could similarly think of conscious feelings as locations in a multidimensional space of phenomenological complexity.

2.7 Conclusion, Objections, and Further Directions

My goal in this chapter was to argue that any biological theory of consciousness must make sense of the full diversity and complexity of consciousness in nature, and this includes the subjective experience of non-human animals. For this purpose, I have introduced my notion of "phenomenological complexity" as something that should be at the very heart of our theorizing about consciousness.

Against those who might be sceptical of the very viability of such a comparative project across the animal tree of life, I have reviewed and expanded the recent call by Birch et al. (2020) to develop a multidimensional framework for the study of animal consciousness, in order to show that phenomenological complexity can be operationalized and in principle be measured by distinguishing five dimensions of consciousness. We have finally developed a battery of tests to probe the experiences of other animals, rather than merely ask where to draw a line in the tree of life. This can be seen as great progress, even in the face of much uncertainty.

Cognitive ethologists such as Griffin were treated with much hostility, but they were right to call out the mistake in identifying a current lack of methods with the impossibility of developing them. Due to the lack of effective methods for probing the subjective experiences of non-human animals, Darwinian theorizing about the function, presence, and origin of mind has long been viewed with much suspicion. But as the present chapter has hopefully illustrated, we have reached a significant stage for a phase transition in which we can use the developed range of experimental paradigms for assessing the minds of other animals as a useful intermediate step for the removal of humans from the centre of reference.

Nevertheless, a multidimensional framework poses new challenges, such as how many dimensions we should distinguish. An objection to the advancement of this project could certainly challenge the five dimensions discussed here. Notably, Birch et al. recognize that their dimensions, such as diachronic experience and self-consciousness, may be correlated – especially once we reach the more sophisticated levels of each – but maintain that they are conceptually distinct. Both the worry and their response are misplaced, however. There is no need to have sharp dividing lines between these dimensions, as long as they offer a superior model to those offered in the past. To assert that their division constitutes genuine conceptual distinctions is beside the point. The conceptual playing field can be carved up in any number of ways, and we should be careful to not mistake such intuitively attractive distinctions for necessary insights into the nature of minds.

The mere conceivability of the possibility of animals with a "richly temporally integrated stream of experiences without any awareness of itself as the subject of those experiences" and others with "temporally fragmented 'staccato' experiences while being aware of itself as the subject of those fragment" (Birch et al., 2020, pp. 798–799) is by itself no more interesting than the assertions by Chalmers and colleagues that we could conceive of zombies that are exactly like us but lack anything it's like to be them.[6] What is needed is an evolutionary approach that asks what the adaptive value of these varying degrees and varieties of consciousness could be in order to assess their plausibility.

The five dimensions should thus primarily be seen as a useful way of overcoming current one-dimensional thinking. This is why I have not set out to fundamentally challenge their proposal. There is nothing to gain from trying to determine the right dimensions *now*. This must be an outcome of future comparative inquiries into the nature of mind and I expect that a greater appreciation for the phenomenological complexity in nature will force us to give up on the idea that such profiles are anything more than useful models to highlight the diversity of minds. There is no such thing as *the* "appropriate grain of analysis" (Birch et al., 2020, p. 797).

Different purposes, such as which species we are comparing, will lend themselves to alternative ways of carving up the differences among them and it is these comparisons that should really matter for a cognitive ethology.

Lastly, it is precisely because of this that my review of the five dimensions began with selfhood, since this dimension most elegantly demonstrates the need for and promise of a strongly gradual and multidimensional approach to the dimensions of animal consciousness. Notably, de Waal (2019) asks us to consider the possibility that "self-awareness develops like an onion, building layer upon layer, rather than appearing all at once" (p. 6). Some may object that such a complex nested view of a scientific phenomenon would make research much more difficult, but if consciousness is a varied and complex phenomenon in nature, we will simply have to learn to change our experimental paradigms accordingly.

Indeed, in the next chapter, I will take de Waal's metaphor of an onion to heart and argue that evaluative experience is the most promising candidate for the origins of consciousness, by shedding the other dimensions aside one by one, in a reverse order from which I argue they plausibly emerged as something like "outer layers" of sentience. This reverse-engineering approach will allow us to significantly reduce the explanatory gap by asking for the most minimal form of consciousness.

Notes

1 Recently, further compelling evidence for mirror self-recognition in the cleaner wrasse has been presented in a follow-up study by Kohda et al. (2022), but it has not yet received critical responses.
2 Another sensory modality worth considering is olfaction. Cazzolla Gatti et al. (2021), for instance, have recently used a sniff test with wolves to demonstrate that they can engage in self-recognition based on smell.
3 Schukraft (2020) offers a critical discussion of whether critical flicker fusion frequency tracks the subjective experience of time.
4 See Danckert et al. (2021) for a useful recent review.
5 I thank Paul Griffiths for this point.
6 See Kirk (2021) for an excellent overview of this debate.

3

THE ORIGINS OF CONSCIOUSNESS OR THE WAR OF THE FIVE DIMENSIONS

3.1 Introduction

We are now in the possession of a wide range of experimental paradigms for an empirical investigation of phenomenological complexity across the animal kingdom. Furthermore, they put us in a position in which plausible sketches, hypotheses, and theories about the evolutionary origins and function of consciousness can and ought to be developed in order to transition towards a true Darwinian science of consciousness.

Curiously, Romanes (1883) maintained that it would be almost impossible to conceive of the dawn of subjectivity without at least a minimal sense of evaluation, going so far as to claim that no more justification than this would be needed:

> [I]f we contemplate the subject, we shall find it difficult or impossible to imagine a form of consciousness, however dim, which does not present, in a correspondingly undeveloped condition, the capacity of preferring some of its states to others – that is, of feeling a distinction between quiescence and vague discomfort, which, with a larger accession of the mind-element, grows into the vivid contrast between a Pleasure and a Pain. I think, therefore, it is needless to say more in justification of the level on the diagram at which I have written these words.
>
> *George John Romanes (1883, p. 111)*

While I share Romanes' view on the origins of consciousness, a modern twenty-first century account of the evolution of consciousness ought to do better than just appeal to one man's intuitions. Other possibilities need to be considered, critically evaluated, and argued against. The previous chapter gave us just these alternatives: five contenders for the crown of the most ancient kind of subjective experience. Who is the oldest in the line of succession and deserves to sit at the centre of our theory of phenomenological complexity?

By "eliminating" one dimension after another from our rich human conception of consciousness, this chapter aims to establish hedonic evaluation as the most promising candidate for an investigation into the origins of consciousness. This Darwinian reverse-engineering approach to the problem of consciousness will allow us to turn the human-centric methodology upside down by asking for its most humble beginnings.

DOI: 10.4324/9781003321729-3

3.1.1 *Chapter Outline*

This chapter is structured as follows. In Section 3.2 "Five Options for the Origins of Consciousness", I narrow down our list of five contenders by arguing that both diachronic and synchronic unity are structural latecomer features of conscious experience unlikely to have been present at the earliest origins of subjective experience. In Section 3.3 "Down to Three", I take on the major challenge of this chapter to make a case against both strongly internalist and strongly externalist views of consciousness that emphasize self-experience and sensory experience, respectively, as the most basic kinds of subjective experience. In Section 3.4 "The Last Dimension Standing: Evaluative Experience", I defend the idea that we should turn evaluation into *the* model for consciousness as a promising teleonomic alternative to a false dilemma between internalist and externalist theories of consciousness. Moreover, I argue that this dimension offers us a promising narrowing of the explanatory gap between matter and mind. Finally, Section 3.5 "The Spoils of War" summarizes my case for the search for the origins of consciousness in the evaluation and respond to some further objections.

3.2 Five Options for the Origins of Consciousness

My treatment of consciousness as a complex multidimensional phenomenon is intended to narrow the explanatory gap by no longer requiring an explanation that demands all of the features of consciousness to appear together as a "one-package deal". A new problem, however, will inevitably arise in the sense that we are apparently faced with at least five different functions and origin stories for the evolution of experience, corresponding to each of the dimensions. Could self-consciousness, synchronic experience, diachronic experience, sensory experience, and evaluative experience all have their own independent origins? Or do some appear first, with others "built" on top? Depending on which dimension(s) we consider to be the most basic, we may be presented with different views on the place of mind in nature perhaps even more diverse than the results from theorizing about the function of the rich human consciousness we are all familiar with.

This is a bullet we ought to bite. In thinking about the function of consciousness, it would be a mistake to think that consciousness only does *one* thing. It has been an unfortunate development that much of the philosophical debate on functions has treated them in a binary and monist fashion (see Matthewson, 2020). Traits are said to have either one function *or* another, similar to how consciousness is seen as either present or absent, without allowing for gradations and variations.[1] But traits can be shaped by numerous selection pressures and thus have a variety of functions that can be realized to various degrees.

It is therefore unsurprising that our complex human experience may not appear to be capturable with a single function statement. Indeed, it is not uncommon among naturalists such as Dennett (2005) to deny that consciousness has a function over and above the cognitive mechanisms constituting it. But such thinking already implicitly presumes a model of human consciousness and complexity. In thinking about the *role* of consciousness in nature, we must address its most humble origins. And as I shall now argue, we can at least make this problem quite a bit smaller by "eliminating" two of the five dimensions that I grouped together under the *unity* of experience, i.e. diachronic and synchronic unity.

3.2.1 *Diachronic Unity of Experience*

Out of all the five dimensions, diachronic unity appears the most readily conceivable as a later feature; a layer of consciousness that may have significantly reshaped experience, but did not give rise to it. As I shall argue in this section, we should understand temporal unity as an evolved feature of the way consciousness came to be structured, not something that is necessary for experiencing itself or would somehow make consciousness "richer".

That diachronic unity isn't required for consciousness may seem the most obvious when we consider *mental time travel*. Unless one buys into a picture in which consciousness can *only* play a functional role if it is connected to episodic memory or, for that matter, the mental simulation of the future – this dimension appears to be strictly optional. This is not to deny that episodic memory has important functions, but that we shouldn't assert the necessity of its presence for consciousness to be functional. Here the burden of proof ought to lie with those who would want to assert that it must be present for the existence of subjective experience. After all, studies on *Severely Deficient Autobiographical Memory* (SDAM) – a condition where those with the most extreme cases completely lack the ability to *relive* their memories – show that there are humans who have subjective experience without this longer timescale form of diachronic unity (Palombo et al., 2015; Watkins, 2018). More, however, has to be said for very short timescales of <5 s, or what I have called the *experience of time*.

James' (1890) description of consciousness as something like a continuous stream or flow of experience akin to a river has been influential, but it was based on human introspection, which is precisely what we seek to overcome. Admittedly, this view has empirical support. The colour-phi illusion experiments mentioned in the last chapter are suggestive that consciousness is inherently dynamic in humans. Yet, what appears like continuous movement to a human may look like a fragmented set of pictures to a different animal, and what in turn looks like a series of snapshots to us may well look like continuous movement to a different species.

From a Darwinian point of view, we should thus see diachronic unity as a gradual rather than binary matter, i.e. temporal experience can be more or less fragmented. Indeed, many animals are likely going to have a higher consciously experienced "frame rate" than humans, which would make their experience less unified, but not inherently inferior. As Broom (2014) has once argued, we shouldn't underestimate smaller animals since "hummingbirds and mice seem to live at a much faster pace than larger, slower-moving animals such as humans" (p. 118). To support his claim, he draws on the critical flicker fusion frequency tests that I introduced in Chapter 2, which have been used by Healy et al. (2013) to demonstrate that larger animals and those with slower metabolisms have a robustly lower temporal resolution. Naturally, a teleonomic view of mind and life should treat the subjective experience of time, the information processing resolution of time, and the speed of an organism's behaviour as tightly linked. Smaller animals with a higher temporal resolution and faster reaction times could plausibly have the advantage of discriminating more individual frames per second than humans can, before it "blends" into motion for them.

If animals experienced time in a more fragmented way, this might constitute an evolutionary advantage in the daily decision problems of such animals, one that just wouldn't pay off in larger animals with slower metabolic rates and decision-making. It is entirely unclear why we should consider a blending of distinct stimuli into a temporal stream of experience as any more "rich" or "conscious" than the blending of my visual experience when I take my glasses off. We may even treat the identification of integration with richness as leading to an *argumentum ad absurdum*, since we would surely not want to accept that leatherback sea turtles (*Dermochelys coriacea*) have a richer temporal consciousness than that of humans or dogs because their flicker fusion frequency is lower and hence suggestive of a higher temporal integration.[2] Instead, a higher temporal resolution or "fragmentation" may very well be seen as similar to higher-resolution vision. Nevertheless, there are plausible adaptive trade-offs between a higher "frame-rate" allowing for more information being gathered per second, whereas a "flowing", more unified mode of experiencing might enable a better understanding of causal processes. But the discussion here should cast doubt on the idea that flow-like experience must be functionally "superior" to a more fragmented kind of experience or that such integration must have been present at the origins of consciousness.

The most reasonable take on diachronic unity would thus be to understand it as a feature of how subjective experience can be structured, something that only comes to evolve *after* consciousness has already made substantial gains in complexity. Metaphors can often be misleading, but they play a crucial role within science (Veit & Ney, 2021), and we can use the common light-switch metaphor to our advantage here: the "lights" at the dawn of animal consciousness might "blink" on and off, only later becoming integrated into something like a stream, without thereby giving rise to an additional explanatory gap.

To accept this evolutionary reasoning is in no way to deny that the way we experience time may be crucial to what it is like to be a typical human subject and agent with long-term goals and a personal identity that is constituted by one's experiences. But I am here interested in the evolution of subjectivity and not (merely) its supposedly most rich and complex form. If we no longer require an account of the origins of minimal consciousness to contain diachronic unity, the explanatory gap has at least been partially reduced, which is all the more reason to be open to a gradualist perspective. The first sparks of subjective experience can plausibly be conceived as something more like fragmented staccato experiences without thereby making them dysfunctional. Let us thus quickly turn to the next dimension of unity.

3.2.2 Synchronic Unity of Experience

Whereas diachronic unity can be readily dismissed as a structural property that may feature in the arrangement of an organism's subjective experience, synchronic unity has long been seen as one of the most fundamental features of consciousness. Indeed, many will have trouble even thinking about the idea of a disunified experience.

In our human experience, smells, tastes, images, pains, etc. simply enter into a coherent "sphere" or "perceptual world" of consciousness: we're living in a field, constructing an *Umwelt*. Everything that "enters" into consciousness appears in your point of view. To even ask the question of whether there is a subjective viewpoint appears to ask for something like an integrated picture of the world and thus presupposes synchronic unity. Again, however, we ought to be careful not to confuse an apparent feature of human experience for a necessary property of consciousness as a phenomenon in nature.

It is important to emphasize that the very terms that have established themselves for discussions about human consciousness may hinder our thinking about other animals. The "subjective" in "subjective experience" may be an unhelpful qualifier for minimal kinds of experience that plausibly do not come in the shape of a subjective point of view. If we consider the possibility of organisms with fragmented experiences, it becomes all too tempting to think of them as multiple little selves or subjects – a view that seems implausible and hence leads to rejecting the possibility of disunified experience. But if subjectivity is something that only gradually came into existence, this possibility should not be rejected merely from introspection into our own *subjective* experience. To further explore this claim, it will be useful to examine Giulio Tononi's *integrated information theory* (IIT), which is the paradigmatic case of a theory that bets on synchronic integration as the core of consciousness and thus serves as a rival to my "bet" on evaluation.

3.2.2.1 Integrated Information Theory of Consciousness

Over the last decades, IIT has been championed by Tononi and his collaborators[3] as an ontological answer to the question of what consciousness is. Drawing their inspiration from William James, Tononi and Edelman (1998) have motivated IIT by maintaining that the "fundamental aspect of the stream of consciousness" is that it is "highly unified or integrated" (p. 1846). Indeed, they go on to assert that it is impossible to conceive "of a conscious scene that is not integrated, that is, one which is not experienced from a single point of view" (Tononi & Edelman, 1998, p. 1846). Instead of thinking about unity as an evolutionary "achievement" that links diffuse phenomenological states such as a pain in one's foot and the experience of red, unity itself becomes the alleged "fundamental property" of phenomenological experience. But is this picture plausible?

If we accept this proposed link between integration and consciousness, the following identity claim of IIT is also credible: "consciousness is one and the same thing as integrated information" (Tononi, 2008, p. 232). This value is then denoted with the Greek letter for phi (Φ) as an information-theoretic measure, or measurement of a system's capacity to integrate information. There are at least two ways to argue against this link between integration and consciousness: i) show that the underlying motivation is misguided or ii) show that a theory based on this link fails to address the puzzles of consciousness.

Nevertheless, in evaluating IIT as a competitor to the pathological complexity thesis, we can readily grant that Φ as a measure of consciousness could have obvious virtues such as the *in-principle* applicability to systems very different from ourselves, allowing them to be placed along a continuum from more to less conscious (Tononi & Koch, 2015). Furthermore, the idea of beginning with minimal theoretical commitments and a very simple model is familiar from other sciences studying complex phenomena. Not unlike the pathological complexity thesis, IIT may thus appear to offer a good starting point for an evolutionary approach to consciousness. Yet, while sharing some of the benefits of the pathological complexity thesis, IIT also has a set of distinct disadvantages. As I shall now argue, while information integration may be very important for consciousness, that would be a far cry from justifying the stronger claims that experience must be unified or that all of consciousness can be reduced to information integration.

Problem 1: The Axiom of Exclusion. A core problem of IIT is its overemphasis on intuitions from human consciousness. This is nicely emphasized in how they deal with the problem that an identification of consciousness with integrated information would, as Merker et al. (2021) point out in a target article criticizing IIT, "entail the coexistence of multiple consciousnesses in a single system at a given time, rather than a single, unified consciousness, as phenomenology and parsimony considerations would seem to dictate" (p. 3). But proponents of IIT simply reject this "possibility of multiple consciousnesses in a single system by fiat", stipulating that only the main complex, i.e. the subset in the system with non-zero Φ^{MAX} (the highest Φ), is conscious (Merker et al., 2021, p. 3). Rather than taking seriously the possibility of distinct streams of experience arising from their emphasis on integrated information, it is rejected through the mere fact that human consciousness appears to be unified. Indeed, the theory hardly seems to address the problem at all, since it is at least conceivable that it will be a different subset of the brain structures or processes that at different times has the largest Φ. This would imply diachronic disunity, which would strengthen our argument against thinking of this dimension as essential, but it also undermines the motivation by IIT proponents to justify their theory through recourse to the idea of a continuous stream of unified experiences, since the view that multiple selves take turns in being conscious does not really seem to be a defence of synchronic unity at all. This "exclusion *postulate*" leads, as Merker et al. (2022) nicely demonstrate in their work, to "some of the most metaphysically bizarre consequences of anything in consciousness theory" (p. 58).[4]

Unfortunately, this is not the only instance where it seems that IIT merely postulates properties of consciousness that allegedly must be "true of every conceivable experience" (Tononi et al., 2022, p. 44), because of introspection into *our* human experience.

Problem 2: The Indicator Validation Problem. By linking **Φ** with consciousness, IIT appears to be able to tell us which systems are conscious and which aren't. Yet, the theory does not appear to have a means for testing that assumption. By this, I am not referring to the common criticism that Φ^{MAX} is computationally intractable and inapplicable to real brains or that the proxy measures of **Φ** deliver inconsistent predictions (see Merker et al., 2021). As I have argued elsewhere with Heather Browning, IIT suffers from what we called an *indicator validation problem*, since "any tests attempting to link **Φ** to consciousness would require prior knowledge of which systems are and are not conscious" (Browning & Veit, 2020d, p. 102). As I shall show in Chapter 4, the pathological complexity thesis does not suffer from this problem since it makes a functional rather than ontological connection between complexity and consciousness, one that does not rely on the human case. It asks: what kind of complexity would make consciousness worth having? If we find systems that have high pathological complexity, yet do not have an evaluative system, that would count as striking evidence against my arguments. IIT cannot make a similar move and instead simply asserts that any system with a non-zero Φ^{MAX} value is conscious.

Problem 3: The Bottleneck of Evolutionary Theory. Lastly, I will mention a problem with IIT that did not receive attention from Merker et al. (2021), leading me to publish the in-print debut of the pathological complexity thesis in a commentary to their target article (see Veit, 2022b). In this commentary – titled "Consciousness, Complexity, and Evolution" – in reference to Tononi and Edelman's (1998) paper "Consciousness and Complexity" – I did not argue as other authors have done that IIT fails for a variety of phenomenological, formal, and neuroscientific reasons, but rather that it fails to pass the bottleneck of evolutionary theory. Admittedly, unity may well be a feature that matters in some way or another, but there is hardly any reason to think that it must be the *only* thing that matters; that they are one and the same. There appears to be an explanatory leftover. As Dennett (2019b) also notes: we must answer what consciousness does for the organism and this is a teleonomic question, not an ontological one. While Tononi (2004) states that his theory would suggest that consciousness "may have evolved precisely because it is identical with the ability to integrate a lot of information in a short period of time. If such information is about the environment, the implication is that, the more an animal is conscious, the larger the number of variables it can take into account jointly to guide its behavior", and this does not in fact provide an answer to the functional question. Less integrated ways of organizing a brain may well help an organism to better guide its behaviour and IIT does not answer how consciousness itself helps the organism. A theory in which integration has, as Godfrey-Smith (2020a) nicely puts it, "no essential connection to sensing, action, and so on" (p. 209) appears biologically misguided. An evolutionary investigation of consciousness requires a sense of complexity that is linked to action, since it is here that consciousness ultimately "pays off". Without answering how unity helps organisms, we cannot make sense of why experience should be unified in the first place. IIT pays little attention to the most important features a biological account of consciousness should offer and thus leaves the phenomenon entirely mysterious.

Verdict: None of these problems provide a "proof" that this theory is fundamentally on the wrong track. My goal here was only to highlight some of IIT's problems that constitute the best illustrations for why we should not just axiomatically assert that unity must be a feature of consciousness. I have argued that a theory that makes unity the fundamental property of consciousness appears to leave out some of the most important features that a theory of consciousness should have. Let us now consider some further reasons why unity may not be all that fundamental.

3.2.2.2 Challenges to the Unity of Consciousness

In the previous chapter, I mentioned that split-brain patients are often used as an example to contemplate the possibility of de-integrated experience – with patients perhaps even constituting two conscious selves (see also Hill, 2018). In a biological materialist view of consciousness, we should expect differences in the organization of the nervous system and information-processing to reflect differences in consciousness. So this line of thinking should be taken seriously to avoid a fallback into dualism, rather than be rejected from the armchair.

In asking for the function of synchronic unity, such studies of pathological cases of consciousness will doubtless be helpful and they have hinted at its role in meta-control. But if we want to learn about the possible advantages of a disunified experience, pathological cases *in humans* may be of little help, since our brain evolved to have a strikingly unified experience. Robbing it of this ability will doubtless lead to many dysfunctions, even if we can learn about surprising abilities that can still be performed with a split-brain. We could, for instance, maintain that it is artificially possible to create disunified streams of experience, while still maintaining that unity is necessary for consciousness to function well. This would no longer be an ontological unity thesis about consciousness, but a functionalist one. While the above arguments may have decreased our credence for thinking that unity must be present for consciousness, they do not necessarily undermine a teleonomic claim about the unity of consciousness. For this, it will be important to take a comparative approach and find natural experiments in the animal kingdom to study *healthy* cases of animals that could constitute natural experiments of organisms resembling split-brain patients.

Citing work by Güntürkün and Bugnyar (2016), Birch et al. suggest that we could rely on birds as an example of such natural split-brain patients since they "have no structure akin to the corpus callosum connecting the two hemispheres of the dorsal pallium, which is homologous to the cortex in mammals" (2020, p. 793) – a structure that was long considered necessary for consciousness. In Chapter 5, I will argue that they overstate their case and that fish and non-avian reptiles make for better examples of disunified streams of experience, since they have even fewer connections between their hemispheres, but birds nevertheless constitute an excellent case for partial unity. Since birds are able to engage in meta-control and complex goal-directed behaviours, not only despite, but because of, their highly lateralized brains, this raises interesting questions about how the division of labour in the brain could relate to the integration of experience – perhaps providing an adaptive benefit.

Challenge 1: Interocular Transfer. Drawing on studies of interocular transfer in pigeons (*Columba livia*) by Ortega et al. (2008), Birch et al. (2020) have argued that they may have limited integration within their vision, which can be divided into two fields: "the red field, which is the lower frontal region important for guiding pecking, and the yellow field, which covers the upper frontal and lateral regions" (pp. 793–794). As Birch et al. highlight, almost all pigeons only succeeded at having interocular transfer between the red fields, thus providing a challenge to the assumption that consciousness must be unified in other animals (see also Hill, 2018). Could such partial unity provide an adaptive advantage? For birds whose eyes are placed on the sides of the head and thus provide limited visual overlap, this may not at all be unreasonable. Each eye might be engaged in a different task, e.g. predator detection versus foraging, which may have an adaptive benefit over the single focus consciousness that appears to demand in our unified experience. Excessive unity in such an animal might even be positively pathological.

Challenge 2: Unihemispheric Sleep. Recall the electroencephalography studies of sleep from Chapter 2, which have demonstrated that some animals let one hemisphere sleep, while the other stays awake. Unlike humans, for instance, aquatic mammals like seals and dolphins appear to have unihemispheric sleep (Mascetti, 2016). Birch et al. (2020) consider this to be further evidence challenging the necessity of unity, since birds have also been shown to engage in unihemispheric sleep. If it is possible to have two streams of consciousness, this could well entail the possibility of dream experiences in one brain, while the other is awake and consciously evaluating things going on around the animal. Obviously, more research will have to be undertaken to answer such questions, but given the presence of a division of labour in brains, we cannot simply assume that consciousness must be a unified experience in order to be functional. Rather than being necessarily pathological, disunified streams of experience may well offer their own adaptive benefits without thereby being "less" conscious. Notably, recent research has suggested that great frigatebirds (*Fregata minor*) are able to sleep in this unihemispheric manner even during flight (Rattenborg et al., 2016). Since flight and swimming contain some similarities, such as the ability to move uninterruptedly in three dimensions for long distances, there might be a case to be made for the convergent evolution of unihemispheric sleep in seals, dolphins, and birds. There is a pathological complexity trade-off between rest and attention that different species appear to solve in different ways.

Challenge 3: Division of Labour and Synesthesia. Lastly, it is worth noting that despite the fact that research on meta-control, crossmodal integration, visuo-spatial bias, and multitasking in animals with lateralized brains may not directly challenge the alleged unity of consciousness, this research can reveal the adaptive benefits of having a brain that is not highly integrated, such as division of labour, and thereby at least indirectly challenge the assumption that unity must be fundamental to consciousness. The most striking cases of lateralization, after all, are found in vertebrates, i.e. in fish and reptiles, that evolved from animals with more "unified" brains. And the above research should at least make us take seriously the possibility that this division of labour is associated with disunified experiencing. In order to think about the experience of such natural split-brains, Hill (2018) offers a useful metaphor of *islands* representing a kind of clustering of particular experiences, "where the components of an island are unified with each other but not with the components of other islands" (p. 4). This metaphor is a good one because it can help us to recognize that even in humans, our islands of experiences may not be all that perfectly integrated. While our visual field very well may constitute one unified sphere – after all this is the sensory dimension that gave rise to the focus on unity – my experiences of sounds, smells, and bodily feelings also appear to be strikingly distinct. They do not perfectly melt together as they might do in someone with *synesthesia*, i.e. subjects who can smell sounds, see shapes for smells, have colours associated with words, and so on (Ward, 2013). It seems hard to deny that the subjective experience of someone with synesthesia shows even higher unification, casting further doubt on the idea that unity is a binary, suggesting instead that it may come in varieties and degrees. Upon reflection, I can perfectly grant that there is a sense in which I have multiple selves – one with a visual point of view, one with goals and motivations, another with a sense of bodily self – that only partially integrate with each other into something like a unified self. That does not appear to be a conceptual error and is quite reasonably necessary for these capacities to play their functional roles. Instead of thinking of human experiences as a wholly unified sphere, it might be better to think of them akin to the partial unity exhibited in the interocular transfer experiments with the visual fields of pigeons. Tentative evidence suggests not only that synesthetes have improved learning and memory (Watson et al., 2014) but also that some suffer from sensory overload and the interference of synesthetic experiences with tasks such as mathematics (Rich et al., 2005), which could help us to understand the trade-offs in having more or less unified experience. Whereas some severe forms of synesthesia are probably pathological, others are plausibly part of normal healthy human variation. Studying these phenomena in more detail will help us to better understand the benefits and drawbacks of unity and put us in a better position to think about how much unity to expect in animals with very different kinds of ecological lifestyles. We should not pretend that unity must be fundamental prior to such empirical investigations.

3.2.3 Unity and Consciousness

Ultimately, this section hoped to demonstrate that the apparent unity of consciousness in humans, whether of the diachronic or synchronic kind, cannot be treated as an insight into necessity regarding what consciousness must be like. In the preface, I mentioned that earlier discussions on the philosophy of mind distinguished between "consciousness" and "qualia" as two distinct problems. It might be worth returning to this older way of thinking, a way that didn't treat the rich properties of human-like consciousness as necessary insights into qualia, which might be much more rudimentary and in principle present in animals very evolutionarily distant from us (see Godfrey-Smith, 2016a, 2016c). Merging these into a single problem may not have constituted progress after all, with the existence of partial unity challenging preconceived notions that consciousness necessarily belongs to discrete subjects. This is why I find the term subjective experience less helpful than have others such as Godfrey-Smith in thinking about the origins and most minimal forms of qualitative experience.[5] Rather than treating "qualia" as a concept that should be analysed through introspection into human consciousness – and thus perhaps inevitably merging these two phenomena before even reaching consensus on how qualia should be conceptualized (see Tye, 2021 for a review) – my goal is to naturalize this notion through an investigation into the most rudimentary beginnings on the evolutionary path towards the complexities of human consciousness. In the eyes of philosophers sitting in the Nagel and Chalmers camp, this might make me an eliminativist, but a better description of this approach would be revisionist, since we revise the notion of qualia in a naturalistically unproblematic manner. As unity does not appear to be necessary for consciousness, we can narrow the explanatory gap by shelving it off from our investigation of the very origins of consciousness.

Indeed, when we consider the simplest forms of subjective experience, unity appears to be in a sense a more trivial affair of definition, rather than an interesting biological property regarding how experience comes to be organized. And these two senses need to be distinguished. If an animal's subjective experiences are limited to a distinction between two senses (e.g. hot and cold, or evaluations of good and bad), there is little point speaking of unity or disunity, simply because there are so few distinct subjective experiences to begin with. Such an animal may have a Φ value high enough to warrant the attribution of consciousness, but it is not unity that does the explanatory work here, since it would be fundamentally in sensing or evaluation that consciousness pays off for the organism. The fact that nervous systems have *some degree* of unity or integration that gives rise to Φ has little to do with consciousness and everything with the fact that all life must show unity at least to some extent. The mere fact that there is "correlation between levels of conscious arousal and phi" (Ginsburg & Jablonka, 2019, p. 126) should therefore not be seen as providing significant support for IIT's insistence that unity is the fundamental property of consciousness.

A passage in Godfrey-Smith (2016d) expresses this quite nicely: "To some degree, unity is inevitable in a living agent: an animal is a whole, a physical object keeping itself alive. But in other ways, unity is optional, an achievement, an invention" (p. 87). And it is these latter, optional, forms of unity that matter for thinking about whether consciousness *must* be unified. In these functional senses of unity, it becomes reasonable to think that the unity of "human experiential profiles might be a later-arriving, derived trait" (2020a, p. 209). We can grant unity may well be a very important feature of human consciousness – perhaps even something that would almost inevitably evolve in systems with heterogeneous washes of qualia – but we should not expect it as a fundamental functional feature of consciousness that explains its raison d'être. From a teleonomic perspective, both dimensions of unity remain of little functional importance until there are animals with a certain degree of phenomenological complexity that allows subjective experience to be organized in interesting new ways to play particular functional roles. Until then, one can hardly speak of integrated organization at all, any more than we should praise an 8-year-old for having a diversified financial portfolio with their two five-dollar notes stored in different places in their room.

Both diachronic unity and synchronic unity appear to be features of the way experiences are organized, not something we need to explain if we are merely concerned about the first sparks of experience in their own right. They do not appear to solve the hard problem of why there are some states that *feel* like something to an organism. While those betting on unity as the fundamental dimension to solve the problems of consciousness may well turn out to be right, it appears more plausible from an evolutionary point of view to treat forms of unity as later "add-ons". Tononi et al. (2022) allege that unity must be seen as a fundamental property of consciousness in any "satisfactory explanation of consciousness" (p. 44), but a satisfactory account of unity can explain it as a special arrangement of the way in which consciousness is organized in humans, without thereby making it a fundamental property of consciousness. Again, we need to be careful not to fall prey to the all-too-common confusion between human consciousness and consciousness as a phenomenon in nature. And since an "elimination" of unity from the necessary properties of consciousness substantially narrows the explanatory gap, this alone would warrant the search for a theory of consciousness in more basic experiences.

3.3 Down to Three

This leaves us with three remaining dimensions of consciousness: i) the experience of a self, ii) sensory experience, and iii) evaluative experience. To think that these must all come together as a one-package deal for consciousness would still leave the explanatory gap incredibly wide. In thinking about the origins of minimal consciousness, it is hence plausible to think that it is in only one of these dimensions that consciousness arose.

Despite the fact that each dimension can provide us with a distinctive model of consciousness, however, the literature has largely focused on the first two options. As I shall argue, this has placed the science of consciousness in an unfortunate dilemma between internalist and externalist approaches to consciousness, i.e. that consciousness is to be explained through recourse of properties either internal to the organism or external to it. While these theories may well succeed at explaining parts of the phenomenon, I will make a case that there remains an explanatory leftover, i.e. precisely what the other dimensions were meant to explain, and thus giving off the impression of an explanatory gap.

My arguments in this section are not intended to rule out or eliminate theories of consciousness focusing on selfhood or sensory experience, but rather to undermine an almost certain confidence seen in parts of the field that it is in these dimensions that the problems of consciousness are resolved. This will help us to motivate looking at an alternative dimension that has the benefits of each dimension, while not sharing their respective weaknesses: the dimension of evaluation.

3.3.1 *Experience of a Self*

Consciousness was once seen as a higher-order form of thought or awareness of oneself as an individual. Moreover, it was treated as something that should distinguish humans from other animals and could perhaps be found in some rudimentary fashion in our close relatives, such as chimpanzees and bonobos, or other intelligent animals such as dolphins and elephants able to recognize themselves in mirrors (Keenan et al., 2003). However, this kind of experience is now more typically regarded as a special kind of subjective experience that is only present in some sentient creatures, the experience of self-consciousness. Indeed, the historical association of the term "consciousness" with some form of rich and exclusively human experience is precisely why many of the researchers engaged in the study of animal consciousness prefer the term "sentience" to refer to more basic kinds of experiences.

Yet, it would be easy to dismiss the dimension of selfhood by appealing to the radical demands of the richest kinds found within its bounds and to contrast these with a simple evaluative feeling. Just as with the dimensions of unity, we should consider the most rudimentary forms within this dimension, i.e. a minimal sense of a bodily self, rather than comparatively rich self-awareness. Here, it will be useful to examine a direct competitor to the pathological complexity thesis that can be found in the *autopoietic* tradition.

3.3.1.1 Autopoiesis and Consciousness

The autopoietic tradition originated with Humberto Maturana and Francisco Varela, who maintained that all "living systems are cognitive systems and living as a process is a process of cognition" (1980, p. 13). A continuity thesis between mind and life can hardly be stronger than this, so it will not be surprising that this tradition has also tried to address the problem of consciousness, perhaps most influentially by Evan Thompson (2007) in his book *Mind in Life: Biology, Phenomenology, and the Sciences of the Mind.*

Maturana and Varela (1980) introduced the term "autopoiesis", which translates as "self-creation", and was intended to encapsulate the organismal phenomena of "viability", "self-maintenance", "self-organization", "self-production", and "self-reproduction". It is these phenomena that they saw being neglected in an alleged object-like view from the Darwinian perspective that neglected the autonomy and subject-like nature of biological organisms. Life, the autopoietic tradition emphasizes, must be bounded and self-maintaining in order to allow its continuing existence and these features should accordingly be the centre of our attention.

Such a view of life emphasizing autonomy and subjecthood certainly lends itself to thinking of life as a bridge between matter and mind. Thompson maintains that consciousness can be described as "a kind of primitively self-aware liveliness or animation of the body" (p. 161), which arises from the "autopoietic identity and sense-making of living beings, but in addition it implies a feeling of self and world" (p. 221). Organisms are seen as being engaged in the "self-production of an inside that also specifies an outside to which it is normatively related" (Thompson, 2007, p. 163). Instead of being passive objects subject to external forces, here they are seen as active subjects determining their own fates with a high degree of *autonomy*. One can readily see why such a framework would lend itself to thinking about the origins of *subjective* experience and why Thompson (2007) frequently invokes Lewontin to argue that the Darwinian/functionalist view of life and mind is mistaken, or at least needs to be supplanted. This underlying motivation of the autopoietic tradition, to provide an internalist alternative to the perceived externalism of Darwinism, is also reflected in related attempts of the tradition to make sense of health, agency, and biological normativity, providing us with a challenge to the pathological complexity thesis on multiple issues of interest.

Granted, it is tempting to base a theory of the organism on these normative notions of vital self-organization. No longer do naturalistic explications of this idea suffer from a failure to distinguish such thermodynamically open systems able to maintain themselves from other systems such as convection cells, hurricanes, flames, and the like. While these also engage in an exchange of energy and matter to maintain themselves away from thermodynamic equilibrium, Nicholson (2014) rightly notes that organisms "are distinctive in that they regulate their interaction with the outside environment by means of physical boundaries (like membranes or skin) that they themselves generate" (p. 355). It is perhaps unsurprising that self-organization accounts have been met with much enthusiasm, with many thinking that they "hold the key to naturalizing rather elusive notions like function, normativity, and agency" (Nicholson, 2014, p. 355). I would happily concur *if* self-organization was the goal of the organism. For systems designed to achieve self-organization as their raison d'être, this would be the best way to naturalize these notions. But living systems have a different rationale.

Problem 1: A Mistaken View of Life. Like the pathological complexity thesis, the work of Thompson (2007) is an attempt to solve the problems of mind within an account of life, ultimately trying to use the properties of life as something like a conceptual bridge to narrow the explanatory gap between matter and mind. But whereas the pathological complexity thesis relies on a Darwinian life-history view of living systems, the autopoietic attempts to provide an alternative to a Darwinian view of life. The problem for such approaches, however, is that the goal of biological systems is ultimately reproduction.

For this, one does not have to deny that "self-maintenance", "self-organization", and "self-production" are real phenomena in biological systems. But they are likewise processes that serve the underlying purpose of the organism, i.e. to participate in a "game of fitness maximization". An adaptationist does not have to demur to the ontological points popular in this literature which claim that "organisms are more accurately conceived as dynamic processes than as stable things" or that the "identity of organisms hinges on the fact that they are continuously undergoing change" due to the fact that they have to deal with thermodynamics (Nicholson, 2014, p. 355). Just because many evolutionary models idealize or "abstract away the temporal dimension of organisms so that they can be studied as static things" does not mean, as Nicholson (2014) seems to suggest, that evolutionary biologists fail to recognize that such idealizations "do not come without costs" (p. 356). This is simply not true. Modern evolutionary theory no longer perceives organisms as mere survival machines. Life cycles are processes, and life history is a central theory of evolutionary biology upon which to build a naturalist Darwinian theory of the organism. As Griffiths (2009) once argued in a paper with the title "In What Sense Does 'Nothing Make Sense Except in the Light of Evolution'?", it is only within the context of evolution by natural selection that we truly understand the teleonomic nature of organisms:

> Non-evolutionary accounts of biological functioning draw the boundary between biological functioning and other, irrelevant causal processes in which organisms are participants in the wrong place. They exclude activities which no-one can seriously doubt are examples of biological functioning. This happens because "viability", "self-reproduction" and the rest are only one component of a more encompassing ability which involves activities whose focus is not on the maintenance of the physiological individual.
>
> *Paul Griffiths (2009, p. 22)*

This quote nicely recapitulates the insights by Millikan that were discussed in Chapter 1: that lots of biological goings-on of organisms are not concerned with self-maintenance but maximizing the number of viable offspring. To this end, the organism will take the risk of undermining its own bodily systems and perhaps even hastening death. One cannot understand the organism – or for that matter biological agency, normativity, and function – without accounting for these Darwinian design trade-offs, or one will positively misunderstand what parts and processes of the organism are *for* and when something goes *wrong*.

Without the Darwinian point of view, we would inevitably end up mischaracterizing the various trade-offs going on in biological design precisely for the real purpose of the organism, which is to maximize its representation in future populations. As Griffiths (2009) once forcefully argued, the mechanisms and processes underlying "big bang mating" strategies "obviously do not contribute to the capacity of individual males to maintain their form. But they do contribute to the life-history strategy by which these males maximize their contributions to future generations" (p. 22). Self-maintenance up until that point is merely a means to an end. If one is interested in the goal-directedness of living systems, one has to understand that it is the organism as an integrated whole that has been shaped by natural selection to contribute to a single goal of fitness maximization, not self-organization.

Problem 2: A Mistaken View of Darwinism. The emphasis on life being inherently dynamic, goal-directed, and normative is something similarly emphasized here as a useful path to reshaping the materialist side of the mind-matter problem. But the autopoietic tradition goes too far in various respects, overreaching in almost the opposite direction from that of the "mainstream" in the philosophy of mind, a fact which owes itself to its origins and conception as a "radical" challenge and alternative to the Darwinian view of life (Escobar, 2012).

It is thus perhaps not surprising that Thompson (2007) explicitly contrasts the autopoietic strategy as an alternative to the Darwinian view of life. In a reply to his commentators, he contrasts the Darwinian reverse-engineering perspective with an autonomy perspective (Thompson, 2011, p. 177). Thompson asserts that the evolutionist's method of reverse engineering is a mere interpretive framework or heuristic and that Darwinians make the mistake of extrapolating from its success as a method to the claim that organisms are design artefacts (2007, p. 211). Instead, he asserts the *autonomy* of life as the fundamental property missed by the functionalist approach. This may be tempting, since the functionalist approach is derided by many as insufficient to account for the basic problem of experience. But the "autonomy-approach", it turns out, is simply empirically inadequate. Owing to its emphasis on an internalist alternative to what are perceived as excessively externalist ways of thinking about life and mind, the approach misses both the fundamentally teleonomic nature of organisms and the reciprocal influence between organism and the world.

Thompson's argument is reminiscent of older vitalist debates and criticisms of evolutionary-functionalist approaches to consciousness for supposedly being unable to capture the most fundamental property of life and mind. It is merely asserted that autonomy is fundamental and sufficient to explain both life and mind. But as I've argued in Chapter 1, the very motivation for conceiving Darwinism (or, for that matter, adaptationism) as a necessarily externalist approach that requires a radical alternative is simply a failure to distinguish the externalist pre-Darwinian design thinking from the more recent teleonomic thinking about design and normativity that conceptually re-engineered them in terms of natural selection and feedback between organism and environment.[6]

Problem 3: A Resistance to Adaptationist Thinking. A life-mind continuity thesis could hardly be stronger than to identify life with a process of cognition. From here, it is only a small step towards a full embrace of biopsychism and to grant consciousness to all life. This would be an odd evolutionary journey, to say the least, but it is being taken seriously by authors such as Thompson (2022). We take one step in the emergence of life and are suddenly presented with consciousness.

Like Godfrey-Smith (2020b), I fear that this tradition makes consciousness too much of "an automatic feature of just being a living organism located in the world" (p. 117), even if its attribution to the most minimal living systems is avoided. Consciousness in such a picture comes in a certain sense for free, which may be intuitively more attractive than a gradualist model, but shares some uncomfortable parallels with epiphenomenalist ideas. It does not seem to give us any purchase on the raison d'être of consciousness. We are not given a gradualist story of how consciousness (or, for that matter, agency) emerges, rather it is something that simply comes *along* with or rather *constitutes* a certain kind of living activity. This makes it hard to think about things gradually becoming more agential, experiential, or – to use their language – "autonomous", precisely because of the resistance to adaptationist thinking.

Indeed, it is hard to make sense of the idea that a vague wash of feeling of "presence in the world" would become refined and enriched through the process of natural selection if these experiences represent neither useful sensory nor evaluative information. The mistaken idea of avoiding the hard problem by trying to rely on a *non-functionalist explanation* makes it ultimately impossible to explain the evolved gradations and variations of subjective experience across the tree of life and this is why we should at least be sceptical that a search for the origins of consciousness in this dimension will succeed. Even if the approach could make sense of the qualitative experience of a self, it is unclear how the other dimensions can be built on a model of self-experience without – again – making them mere automatic features of living activity. We do not appear to be able to answer the question of why some sensory and evaluative processes are felt whereas others are not. The resistance to adaptationist thinking makes us unable to think about the most important properties of consciousness, i.e. what it does *for* organisms, and thus deprives us of a unique explanatory strategy for the life sciences.

Problem 4: A Neglect of Feedback. Lastly, in trying to motivate a view of life as an active and autonomous agency, external goings-on come to be endogenized as just another kind of organismal activity, leading to a neglect of the "to-and-fro traffic characteristic of organism/environment relations" (Godfrey-Smith, 2016b, p. 778). Instead of making the externalist move of seeing internal events and processes as just another kind of environment, they neglect the role of the organism's environment in an effort to not undermine the "autonomy" of the organism. The environment is almost seen as just another kind of internal goings-on, the organism forcing itself onto nature, without reciprocal influence. Thompson (2007) urges us to see autopoietic systems as "sources of their own activity, specifying their own domains of interaction, not as transducers or functions for converting input instructions into output products" (p. 46). A strikingly internalist picture is provided in which organismal properties are said to emerge between top-down and bottom-up processes within the organism, in a manner of supposedly naturalistically unproblematic "circular causality" that has been seen with suspicion by more reductionist-inclined scientists.

Godfrey-Smith (2016b) describes writers in this family of views as being overly zealous in their determination to describe and explain living systems without any reference to passivity. In an anti-externalist mantra, they argue that organisms must be seen as subjects *instead* of objects, neglecting the possibility that agency can come in degrees and that organisms are both subjects *and* objects. Indeed, Thompson (2007) tellingly describes the nature of any autopoietic system as being "defined by its endogenous, self-organizing and self-controlling dynamics" which do "not have inputs and outputs in the usual sense" (p. 43). While recent work on the mechanisms of such self-organizing systems does not deny the interaction between organism and environment – highlighting dynamic feedback, and the need for such systems to maintain organizational and operational and organizational closure in order to stave off thermodynamic processes towards entropy – the emphasis on autonomy has led, as Godfrey-Smith (2016b) notes, to a continuing resistance "to the role of ecology, in a broad sense – resistant to the fact that it is part of the nature of life to be in ongoing interaction with an environment that is *other*" (p. 778). But this neglect of ecological feedback is precisely why the theory appears unable to explain the heterogeneity and adaptive fit of subjective experience to the life histories of animals.

Verdict: Again, none of these problems must spell doom for the autopoietic approach. While their opposition to Darwinism characterized as a necessarily externalist program is mistaken, that doesn't necessarily mean that their internalist approach to the mind is. Nevertheless, the resistance of the approach to adaptationist and functionalist thinking makes it hard to see how they could account for the dimensions of sensory and evaluative experience. If we are interested in a strong continuity thesis between life and mind, it appears that we would be led to life-history theory and the pathological complexity thesis, since an evolutionary approach to life and mind is not failing to consider the autonomy of the organism – instead, it simply shows that autonomy ought not be conceived as the fundamental property of either life or mind. If there is anything like such a fundamental property, it would be the teleonomic trade-offs any Darwinian creature has to deal with in the maximization of their fitness, and it is in the complexity of these trade-offs, I argue, that we find the origins of consciousness.

3.3.1.2 How to Think about Selfhood?

Despite flaws in the autopoietic approach, the basic idea of consciousness being related to life, self-production, and agency is a useful one and has been influential both in Godfrey-Smith's attempts to naturalize subjectivity and Ginsburg and Jablonka's (2019) recent work on the evolution of consciousness. They maintain that consciousness is something like a new mode of being or way of life that has arisen somewhere during the aforementioned Cambrian explosion; a particular animal way of being in which subjective experience is simply part of what it means to live a flexible life. Note that de Waal did not assign a zero level of self-awareness to any animals in his depiction of a gradualist view of self-awareness (**B** in Figure 2.2). This is because he thinks that all animals require at least a minimal form of a self-concept. But treated literally, this claim – despite being intuitive – is surely too strong when we think of the less mobile and more "plant-like" side of the animal branch of life such as corals, anemones, and sponges.

To think that consciousness evolves in organisms that become more recognizably agent-like does not have to mean that a sense of self lies at the origin of consciousness. Self-related capacities might be close to necessary for the lifestyles of animals in the older but still common folk understanding of animals as mobile organisms with a nervous system: a particular animal *mode of being*. Here, a registration of a difference between self and other seems to be a requirement for efficient movement and action selection in a complex body (see also Godfrey-Smith, 2020b). After all, such animals must have some way of distinguishing themselves from the world. As Birch et al. put this point, "[a]ny complex, actively mobile animal needs a way of disentangling changes to its sensory input that are due to its own movements from changes due to events in the world" (p. 797).[7]

But that doesn't necessarily mean that the self itself is experienced. The hard problem challenge appears to hardly be answered here, given that many of these cognitive processes appear to operate unconsciously in humans. Worse, we do not seem to be able to provide an answer as to how sensory and evaluative experiences evolve out of a feeling of selfhood, making this dimension look more like a later invention. The problems of the autopoietic tradition appear to generalize to any internalist view of consciousness that doesn't focus directly on feedback between organism and environment, and the role of selfhood for the organism. This doesn't mean that properties related to selfhood, agency, subjecthood, and the like are unimportant, but that it isn't in the experience of a self *per se* that we find the fundamental property of consciousness.

Consider the evolutionary picture offered by Feinberg and Mallatt (2016) in their *The Ancient Origins of Consciousness*, where a distinction is drawn between exteroceptive, interoceptive, and affective sides of experiences. In thinking about the origins of self-consciousness, it does not appear hard to imagine that it simply constitutes a combination of these capacities. The most minimal form of selfhood would constitute a distinction between exteroception and interoception and this would make the evolution of self-consciousness look like some kind of "disentangling" on the sensory side of consciousness. It should not be expected from the earliest organisms possessing some wash of sensory sensation and hence should be seen as a later layer added on top of this. What we thus need in order to uncover the ancient origins of this subjectivity is a more gradualist picture of that basic idea – in which complex cognitive processes are recognized that, as Godfrey-Smith (2020b) argues, largely go on behind the scenes, unlike the features that we now usually associate with the core of consciousness. The dimension of selfhood is more plausibly seen as something gradually build out of more basic sensory and evaluative experiences, eventually giving rise to a meaningful recognition of a subject and thus *subjective* experience, but it is in the actual building blocks of experience that we must seek the origins of consciousness.

Let us not add to the number of ingredients evolution has to produce in one fell swoop for consciousness to arise, as if (in the example too often used by creationists) the human eye magically appeared into existence. As Ginsburg and Jablonka (2019) point out, Darwin had trouble conceiving of how the eye could gradually appear, "but as he explained, this was the problem of a failure of his imagination rather than a failure of his evolutionary theory" (p. 32). A gradualist picture may be wrong, but those studying consciousness have so far spent little effort on trying to make the best possible case for it.[8] A Darwinian picture, while perhaps hard to think about, may be the only path to finding the true place of mind in nature. We can happily embrace the possibility that there could be something like "hemi-semi-demi-pseudo-proto-quasi-minds" (Dennett, 1995b, p. 108) that may appear in creatures very different from ourselves. From this bottom-up perspective, the construction of a self, of a subjective point of view, can again be seen as an outer layer in the onion that is consciousness. Shedding it away gets us one step closer to finding the origins of subjective experience.

This finally leaves us with two remaining options and a substantial narrowing of the explanatory gap. While diachronic experience, synchronic experience, and the experience of a self can be understood as features *of* the way conscious experience can be structured, as opposed to constituting it, it remains an open question as to whether the origins of consciousness can be found on the sensory or evaluative side. In the following, I will argue that there are reasons to doubt that the answer is to be found in the sensory dimension, which suffers from what are almost the inverse problems from the dimension of selfhood.

3.3.2 Sensory Experience

A sensory explication of phenomenal experience is widespread among both philosophers of mind and neuroscientists investigating consciousness. The sensory side and, in particular, studies on human vision have often served as the model for *all* of consciousness, likely due to its ties to both of the equally influential notions of a *point of view* and an *awareness of*. It has been seen as the key to understanding phenomenological experience, which was only aided by the fact that even a science of human consciousness has long been deemed impossible and continues to be seen by some with suspicion. To a large extent, the focus on vision was an attempt to make research on consciousness as scientific as possible, since this capacity seemed more readily testable.

Affective neuroscience, which focuses on the moods, personality, emotions, motivations, and feelings of both humans and animals, has continued to be viewed with suspicion not only among cognitive neuroscientists who saw this research as too closely related to "consciousness science" (see Cacioppo & Gardner, 1999) but also among philosophers who saw this empirical research as largely irrelevant to their conceptual analysis of emotions (see Griffiths, 1997, 2017).

The problem with this approach, of course, is the possibility that our early models and theories, developed based on human vision, may influence the way we treat all other aspects of subjective experience. Perhaps our current troubles with naturalizing consciousness are due to path dependence and we would have been in a much better place if we had begun with the study of affect and valence (see also Solms, 2021). But to advance such an alternative approach, I will first argue against the centrality of a sensory model of all experiences.

3.3.2.1 Experimental Philosophy of Consciousness

One source of evidence against a sensory-centric view has so far been given little attention in the philosophy of mind and that is experimental philosophy (of mind). In particular, recent work by experimental philosophers such as Justin Sytsma and Edouard Machery on the intuitions of the public regarding their conscious experiences appears to offer a strikingly different picture from the major views within philosophy of mind and one that fits strongly with the evaluation-first view defended here (Machery & Sytsma, 2011; Sytsma, 2010, 2012; Sytsma & Machery, 2009, 2010, 2012; Sytsma & Ozdemir, 2019).

In an influential study, Sytsma and Machery (2010) show that members of the public do not share the philosophical consensus view that consciousness is characterized by its "phenomenality". While there is much debate on how this notion ought to be conceived, there is broad consensus in the literature that "subjectively experienced mental states have phenomenal properties: There is something it is like to see red, smell banana, feel anger, and be in pain" (Sytsma & Machery, 2010, p. 324). Ordinary folks, however, do not appear to recognize a commonality between all of these states. When asked whether a relatively simple robot can feel pain, ordinary people are highly sceptical. However, when asked whether the robot can see red, the folk were much more likely to say yes compared to philosophers in the same study. This suggests that the folk does not share the philosophical conception of consciousness and subjective experience, i.e. they do not define consciousness as philosophers do, in relation to their "felt" or "phenomenal" properties (Machery & Sytsma, 2011). But if they do not rely on the philosophers' "consensus" view, this raises the question of how they do think about consciousness.

The findings of Sytsma and Machery suggest that the ascription of conscious mental states to a robot rests on a hedonic evaluation, i.e. whether something feels good or bad. Ordinary people appeared to distinguish those states with valence or affect such as pain, moods, and emotions and those without such as a pure sensation of red or the smell of isoamyl acetate (Sytsma & Machery, 2010). Ordinary people seem to perceive consciousness as an evaluative experience, thus refusing to attribute mental states to robots when they include experiences that we would typically associate with hedonic evaluation. This would be a radical inversion of the way philosophers have thought about the problem, with evaluation here taking much more centre stage.

Mere sensory discrimination did not appear to be particularly controversial when attributed to a robot, which is strikingly different from Dennett's (1996) assertion that the robot Cog[3] "cannot yet see or hear or feel at all" (p. 16) (cited in Sytsma & Machery, 2010, p. 302). And yet, ordinary people seemed resistant to assign a robot the ability to perceive "familiar smells associated with either positive or negative valence" (Sytsma & Machery, 2010, p. 318). The common folk concept of subjective experience thus appears to group "different types of perceptual experiences, bodily sensations, and felt emotions depending on their valence" such that only those sensory processes linked with valence are considered restricted to conscious beings (p. 318). These results are interesting because they may undermine the very foundation for the philosophical resistance of those defending the idea that there is something like an unbridgeable "explanatory gap" (Levine, 1983) or a "hard problem of consciousness" (Chalmers, 1995):

> The hard problem is typically justified on the grounds that we are acquainted with the phenomenal properties of states such as pain and seeing red and that functional accounts of mental states fail to explain how they can have such phenomenal properties. Our findings challenge the first premise of this argument. Because people do not seem to conceptualize their subjective mental life as phenomenal, it is at least unclear that we are pretheoretically acquainted with the phenomenal properties of our conscious mental states.
>
> *Sytsma and Machery (2010, p. 324)*

Philosophers have long religiously maintained that subjective experience is essentially conceived in the same way by both themselves and the public – that they are merely using the folk concept of consciousness. Sytsma and Machery (2010) list multiple high-profile examples of this widespread belief even among naturalist philosophers (p. 320), which I shall repeat here: Alvin Goldman (1993) comes out in support of "the basic integrity of the folk-psychological conception of consciousness and its importance in cognitive theorizing" (p. 364). Ned Block (2004), in his "Qualia" entry in *The Oxford Companion to the Mind*, similarly defends the phenomenological view of consciousness among philosophers as *the* folk view. Patricia Churchland (1988) also accepts that the public hold such a view, but urges us to consider the "outright replacement of the old folk notion of consciousness with new and better large-scale concepts" (p. 302). Even Dennett (2005), who likewise wants to develop a scientific account of consciousness by more or less banishing the notion of "qualia", holds that it is part of our folk conception of consciousness (p. 27). Moreover, the study of Sytsma and Machery (2010) also asked philosophers to make a prediction of the assessments ordinary people would make, which they largely expected to be analogous to their own answers (though slightly less sceptical). But as Dennett's assessment of Cog[3] indicates, philosophers may have been strikingly mistaken by confusing the views of their discipline with that of the public.

If this way of thinking about consciousness, however, rests in a mere artefact of philosophical training, we may have to radically revise our assessment of the common philosophical critique by the likes of Chalmers and Nagel, who have maintained that neuroscientists and psychologists who claimed to have explained consciousness have naively failed to address the "obvious" hard problem of consciousness.[9] If these experimental results are indicative of the way non-philosophers think about the problem of consciousness, such assertions ought to be seen as quite an uncharitable interpretation of what the scientists are doing and thinking.

Sytsma and Machery (2010) posit that "it might be that like the folk, they do not conceive of subjective experience as being phenomenal, in spite of having plausibly carefully considered [']what it is like['] for them to see red, feel pain, and so on" (p. 323). Both the public and scientists may simply not consider a further phenomenological property or "qualia" that needs to be addressed, which they think is in line with their own experiences, where "many ordinary people either don't understand or don't take seriously the philosophical concept of phenomenal consciousness even after a lengthy explanation" (p. 323). If the hard problem is only something to be recognized once one has come to be "indoctrinated" by the orthodoxy in the philosophy of mind, this may explain why some scientists, after passionately endorsing the possibility of and need for a science of consciousness, have later taken on the view that the hard problem cannot or may not yet be solved and that they are instead focusing on the "easy" problems of consciousness (see also Crick & Koch, 1990; Klein & Barron, 2020). This may in some cases be merely a strategic choice to avoid the charges of naivety hurled against those scientists claiming to provide explanatory sketches and theories of consciousness. But if so, it would be an unfortunate one since it in turn legitimizes talk of the hard problem as something that cannot be overcome by ordinary science.

Once a naturalistically inclined scientist interested in developing a science of the mind gets too close to what Dennett (2017a) described as the "Cartesian gravity" of Descartes' dualistic way of thinking, it becomes all but impossible to escape once one has orbited too close to "Planet Descartes" (p. 20). But this is what the very mention of the hard problem unfortunately enables. If the vision-centric bias in our thinking about and research of consciousness makes it all but impossible to shake off the dualist leftover of what Dennett (1991) described as a "Cartesian theater" in which everything is consciously presented to a homunculus, we may wish to return to an earlier evolutionary view suggested by Dawkins (1998) in which "[t]he key to the origin of consciousness itself may lie in the emotional experience of suffering" (p. 324). This at least is the core of the pathological complexity thesis defended here: evaluative experience as the original and most basic kind of subjective experience.

Ultimately, Sytsma and Machery (2010) have provided us with a beautiful case for the usefulness and perhaps even necessity of experimental philosophy to progress in philosophical debates muddled by appeals to intuition.[10] Godfrey-Smith (2020b), while accepting that "Sytsma and Machery may be right about the everyday conception of experience", nevertheless maintains that "everyday thinking may also be mistaken" (p. 311). That is certainly correct. However, in a debate that has extensively drawn upon near-certain assertions about the folk concept of phenomenality as a support on which to build our models of consciousness located on the sensory side, this work at least undermines a central argument for betting on this dimension.

Due to considerations of space, I will not offer a detailed defence of why experimental philosophy is useful to philosophers. Firstly, I have done so elsewhere (Browning & Veit, 2022b; Veit, 2021c), and secondly, so have many others (see Knobe & Nichols, 2017 for an overview). Within the context of this book, it is sufficient to acknowledge that the intuitions of the public may not settle this debate, but that we should at least take seriously the insight of the folk into subjective experience as a source of evidence in judging which dimension of consciousness should be seen as the most fundamental. Let us now turn to my core argument against the sensory side, i.e. an excessive externalist representationalism.

3.3.2.2 The Problems with Externalist Representationalism

Yet another elegant argument against the centrality of a sensory model has come from Godfrey-Smith (2020b), who despite his insistence on a possible separation between the two dimensions of sensory discrimination and evaluation has argued that a "problem with much recent work in philosophy is the idea that sensing is not only an important part of experience, but just about all that goes on there" (2020b, p. 113). Indeed, much of the recent work in the philosophy of mind uses the words "sensing" or "perception" as umbrella terms to cover all forms of subjective experience. Moods, emotions, and feelings are seen through this lens as the detection of some internal phenomena such as thirst or hunger. The only difference of these from olfaction and vision is a matter of perceptual direction: inward versus outward. Godfrey-Smith (2020b) uses two influential examples from different generations: Fred Dretske, who influenced Godfrey-Smith's own representationalist views when he was a student, and Jesse Prinz, who was his colleague at CUNY and in turn appears to have made him sceptical of the promises of representationalism. While Dretske (1993) doesn't assert that all of the subjective experience must be perceptual, he cites Velmans (1991), Humphrey (1992), and the originator of the influential *global workspace theory* Baars (1988) to support the view that perceptual experience and belief are taken to be the "clearest and most compelling" paradigm cases of consciousness in empirical research (p. 272). In a revealing passage, Dretske argues:

> Why can't we, following Damasio (1994), conceive of emotions, feelings, and moods as perception of chemical, hormonal, visceral, and musculoskeletal states of the body?
>
> This way of thinking about pains, itches, tickles, and other bodily sensations puts them in exactly the same category as the experiences we have when we are made perceptually aware of our environment. The only difference is that bodily sensations are the experiences we have of *objects* in the body (the stomach, the head, the joints, etc.), not *objects* outside the body.
>
> *Dretske (1999, p. 117) [italics added for emphasis]*

But it may well be a mistake to think of these experiences as representing *objects* in the body, which makes these accounts feel very much like attempts to somehow objectively experience states of the world as described by Newtonian mechanics. The very reason these accounts haven't satisfied proponents of the hard problem has been the lack of recognition of a *subject*. In order to address these problems, *subjectivity* needs to have a role and not be ignored as a mere by-product of the existence of sensory representation. But this is next to impossible due to a reliance on a strongly externalist mode of explanation. Rather than recognizing the epistemological straightjacket of a strongly externalist approach, representationalists tend to simply bite this bullet and declare the sense of a self or subject *unimportant*.

At the end of his monograph *The Conscious Brain: How Attention Engenders Experience*, in which he defends his Attended Intermediate-Level Representation theory of consciousness, Prinz (2012) confidently asserts that "[a]ll consciousness is perceptual; there is no distinctive cognitive phenomenology or any phenomenal self" despite his own concession that "[almost] all of the empirical research reviewed here comes from vision science" (p. 341). Indeed, his own theory is simply an extension, if not generalization, of his earlier representationalist neurofunctional theory of visual consciousness (see Prinz, 2000), with little to no attention paid to conflicting empirical work on other dimensions of consciousness.

Such unashamed confessions of confidence in the centrality of vision as the paradigm of all of consciousness should at least raise some worries that this path may have been something like a wrong turn. Godfrey-Smith sees this as a general tendency of much current information-theoretic work on consciousness – such as Michael Tye's (1995) PANIC theory of consciousness and Stanislas Dehaene's (2014) version of the Global Workspace Theory due to Baars – treating consciousness as a special qualitative way of information being represented in a mind (Godfrey-Smith, 2020b, p. 115). These representationalist theories share too much with an older empiricist view of the mind "as merely reactive, needing to derive its patterning from elsewhere" (Godfrey-Smith, 2020b, p. 188). That such a view leaves something out was rightly recognized by internalists such as Thompson (2007): "Representationalism neglects the subjective character of experience" (p. 283) and if "there is to be progress in understanding mental imagery as a form of human experience, and not merely as a form of mental representation, then we need to do better" (p. 267). But to do better, we do not need to resort to internalist views of the mind.

3.3.2.3 An Alternative to Internalism and Externalism

Godfrey-Smith (2020b) urges us to introspect on whether sensory experience and belief are really the most compelling cases of consciousness. He suggests that what I have discussed under the label of affects in Chapter 2, i.e. "emotions, willings, moods, and urges", seem to be at least as and in his own case "quite a bit *more* clear, as cases of conscious experience, than beliefs" (p. 114). But even if we were to allow that vision is the most paradigmatic case of human conscious experience, this may simply be an artefact of our evolutionary path, a path that has made us masters at learning about and improving our environments, thus making vision appear to represent the external world "objectively", like a mirror-image presented in something like a Cartesian theater.

Godfrey-Smith suggests that an "alternative to [the strongly representationalist] view, rather obvious but neglected, is the idea discussed in our section on the experience of selves that a mood is not a *presentation of* some fact or condition; it is just *the way things are* with you, at that time" (2020b, p. 114). That was one feature of the selfhood-first view that made it attractive. However, both the autopoietic tradition and the representationalist tradition fail to recognize feedback between internal and external factors. Treatments of consciousness by Prinz, Dretske, and others neglect the organism's dynamic role in subjective experience, simply treating qualia as little more than representations of an environmental state. Internal goings-on such as hunger are seen as just another kind of environment. No place is given to a subject, there is no centre of agency, and there is no feedback between the environment, action, and consciousness, making it unsurprising that such views turn the place of *subjective* experience in nature into something like a mystery. Subjectivity simply slips through the cracks in such a widening of the explanatory gap.

Here, we can follow Godfrey-Smith's suggestion to "reject the un-ecological side" of the autopoietic anti-Darwinian emphasis on autonomy and instead embrace the importance of both input and output, or perhaps more importantly the causal traffic that goes on between both sides and expresses itself in action (2016b, 788). Unlike Godfrey-Smith, however, I argue that this traffic that is important to understanding the function of consciousness is readily explicable from the evaluative side of experience, which is directly tied to action.

Whereas the dimension of sensory experience emphasizes an externalist view of consciousness, the dimensions of self-experience emphasize internalism, two kinds of overreaching that make it extremely hard, if not necessarily unsatisfactory, to explicate either side in terms of the other, which is reminiscent of the dilemma the ethologists faced between the vitalist's internalism and the behaviourist's externalism. Here, evaluation offers us a way out of this dilemma by being inherently dynamic and agential. What makes an external state good or bad depends on the state of the organism, and likewise whether a state of the organism is a good one depends crucially on the environment. Ecological feedback is built into this Darwinian picture of an evaluation-first view from the very beginning, with action and agency being emphasized.

3.4 The Last Dimension Standing: Evaluative Experience

My goal in this chapter was not to provide definite proofs that theories of consciousness centred on the other dimensions must be wrong, but rather to weaken the confidence one may have for thinking of these dimensions as fundamental to consciousness. In peeling off the dimensions that I consider to be later transformations of consciousness rather than its evolutionary origin, it should hardly be surprising that the picture becomes less certain the closer we move to the core of this mysterious onion. The sensory side, in particular, is a dimension that any account that bases the evolutionary origins of consciousness in evaluation must explain as it would otherwise do no better than its contenders. In offering my account of how sentience evolved in Chapter 4, I will also detail a picture of how the representational sensory side of consciousness, as well as selfhood, could have quickly evolved in the Cambrian once a hedonic mode of being was "set in place". Whereas theories of consciousness centred on these dimensions appear to explain the other dimensions somewhat inadequately, I will argue that a theory of consciousness centred on sentience will allow us to keep what is best about both approaches.

After all, many of the experimental paradigms discussed in the previous chapter to distinguish conscious from unconscious perception rely on particular forms of learning such as trace conditioning, which suggests that evaluation plays a core role in consciousness. Nevertheless, before I turn to valence as a resource for an alternative theory of consciousness, we should also critically assess the evaluative dimension as I did with the other dimensions. Two important challenges will be met: to motivate i) that hedonic valence constitutes a common core to the evaluative dimension and ii) that it can help us to make sense of the origins, function, and phenomenological complexity of consciousness.

3.4.1 Hedonic Valence as the Core of Evaluation

What is it that unifies the evaluative states we call moods, feelings, and emotions? As Browning (2020b) readily acknowledges in her attempt to naturalize a subjective notion of animal welfare: what we usually call affective states are "extremely heterogeneous states" (p. 164). However, it appears that not only ordinary people see valence as the common denominator of all these "affective states", but also a growing number of scientists who study them. Indeed, as I mentioned in Chapter 2, many consider valence to constitute a "common currency" that makes all of these heterogeneous states subjective and comparable. While Browning (2020b) defends a Benthamite view of subjective well-being as the common currency of affective states, she considers the possibility that our evaluative experience could be more of a construct, rather than an integrated state: "something like health, made up of multiple different components that, though individually real, do not together form any naturally existing state" (p. 164). I like this comparison because it leads us back to the close connection between health and consciousness that motivated the core thesis of this book. As I have argued in Chapter 1, health is not just a construct – it is an optimal design-response to the species-specific pathological complexity faced by different organisms in their life histories, with fitness providing a common currency of evaluation. Similarly, hedonic valence can be seen as the common currency of evaluation for an agent's choices between alternative actions.

But despite the agreement among many scientists that valence constitutes a common currency in humans and many other animals *now*, this doesn't necessarily mean that such a currency was present at the very origins of subjective experience, nor that such a "currency" must exist. While I will defend both claims in the next chapter, I admit that the notion of a "common currency" in thinking about the very origins of valence can be something of a mixed bag because it seemingly presupposes the existence of multiple distinct subjective experiences that are to be compared. Talk of a currency may incorrectly suggest that valence evolved in order to make the different aforementioned states comparable – similarly to how real monetary currencies arose only after there were already goods that could be traded in a more efficient way through the implementation of a common currency. Valence could then be seen as having evolved in response to a certain kind of phenomenological complexity on the sensory side and this appears to be a popular view among those endorsing strongly externalist representationalist views of consciousness. However, we have seen that valence cannot readily be explicated within a sensory model of consciousness. The origins of valence are better conceived as the origins of fuzzy action imperatives, which arose out of something like the vague discomfort Romanes describes, but which then evolved to have richer discriminatory capacities between different states. Or to put it differently, hedonic valence becomes a common currency through the evolution of richer discrimination and thus the origin of sensory representations. This would explain the evolution of sensory consciousness, the distinction between conscious and unconscious perception within an evolved valence system of the brain, and it seems to fit the folk understanding of consciousness. This nuance aside, however, I consider the term on balance a useful one to understand the origins of consciousness.

An objection to this view would be to insist that even if we think that hedonic valence constitutes a common currency in our decision-making, our own human experience may strikingly differ from other animals, who may lack such a common currency. To this, of course, we can simply respond that pleasure – unlike language or higher-order symbolic thought – does not seem like something restricted to us. Indeed, it may even be more important for animals to possess a common currency of pleasure and pain, since they cannot engage in the same symbolic cognitive processing as us. In humans with a rare pathology making them unable to feel pain, i.e. congenital analgesia (sometimes referred to as congenital insensitivity to pain), early death is common due to a neglect of or inability to detect injuries and diseases (Cox et al., 2006; Nagasako et al., 2003; Thrush, 1973). Importantly, the absence of pain should not be confused with the absence of any negative valence. It is a mistake to think of hedonism as referring to just two states of pleasure and pain; these do not exhaust the scope of all valenced states (whether positive or negative).

The experience of pain nevertheless plays a crucial role in developing a concept of self and one's body in relation to its environment. Those with congenital analgesia must exert cognitive effort to think about or actively represent the potential dangers to their body, since they lack a system of "punishment" that would teach them from their childhood onwards. While this is certainly not easy, and less efficient than actual pain-experience, it can be done. Animals, however, are unlikely to even make it that far without the fast decision-making and learning of important associations that are enabled through the experience of negative valence that can be traded off against other needs such as hunger. They would not be able to think about the likely "harms" of particular behaviours without painful experiences to make these connections. Such evidence is compelling and the next chapter will offer an account of how this capacity may have evolved, but intuitive thinking about hedonic decision-making appears to be at least plausibly on the right track in regard to a potentially much more ancient felt experience.

Browning (2020b) puts the centrality of valence in our subjective experience vividly: "I consider myself right now and the combination of states I am experiencing – mild hunger, physical comfort in my office chair, slight head pain from a lingering cold, anticipation of my upcoming lunch, some intellectual discomfort from trying to write this chapter, among other states" (p. 169). While these appear to be strikingly different kinds of experiences, there does seem to be an intuitive sense in which they become integrated on something like a single scale. They have a felt evaluative sensation that becomes integrated into a single state and enables us to compare competing interests and motivations. This comparison does not take place in the cognitive or representationalist sense of a calculation, but rather an instantaneous general experience of one's state – a total state of momentary *feeling* in just that sense of the word. In his work on consciousness, Damasio (1999) similarly centres feelings and evaluation to make sense of consciousness as promoting advantageous actions, but he requires some minimal sensory representation and feeling of a self at the evolutionary origins of consciousness. I do not see these as necessary for minimal sentience since it gets us uncomfortably close to the idea that consciousness "pops" into existence once there is a certain degree of complexity in self-recognition, sensory processing, and evaluative cognition. Again, such thinking appears to make the qualitative aspect of consciousness something like a mysterious extra ingredient, although his emphasis on embodied cognition is important to make sense of the role of consciousness in nature.

As I shall argue in the next chapter, what I call *Benthamite creatures* have a nexus of evaluation at their final behavioural common path that enables them to engage in efficient decision-making under the conflicts of different needs, even in the absence of a sense of self or felt representations.[11] Valence plausibly came into existence with a basic feeling of good and bad, without any *felt* sensory richness. As long as the first sentient beings continued to engage in a behaviour that caused them something like a "plus sign" and changed it up once it became a "minus" (in the sense of a vague impervious impulse for action, possibly realized through dopamine and other valence-related molecules with a deep evolutionary origin), there would have been no need for the presence of the other dimensions for this to be adaptive. Under very deflationary senses of the term "representation", such affective decision-making may well be considered as representing two different states, "good" and "bad", in the sense of fitness-enhancing (e.g. Carruthers, 2018, in press), but such a usage of the term is unhelpful in understanding i) why some representations are felt, and ii) how consciousness becomes more representational. Importantly, the absence of conscious representations does not imply the absence of representations altogether. Most of what goes on in the brain and nervous system is unconscious and many of these processes may well be representational. But before I explicate this account of mine in the next chapter, let us return to the question of how far we have come in narrowing the explanatory gap.

3.4.2 Valence, Qualia, and the Explanatory Gap

Sytsma and Machery see their results as undermining the very idea that there is *anything* like a hard problem or explanatory gap. I think that deflates the problem too much, though I can see the temptation to adopt this position. Unlike Sytsma and Machery, I see the move towards understanding the biological basis of valence and affects as a *naturalization* of the vexing notion of "qualia" through an alternative non-vision-centric model of consciousness. While Sytsma and Machery raise the possibility that the valence of consciousness is a problem akin to the phenomenological version of the hard problem, they resist the notion that we couldn't explain it by virtue of functional and mechanistic explanations. They consider it to be straightforward to understand that the "hedonic value of a stimulus or a bodily state seems to be an evaluation of its expected value to the organism" (p. 322). There doesn't appear to then be an additional problem of *why* there is valence. Like Solms (2021), I argue that this makes the evaluative side of experience a compelling target for an attempt to bridge the gap between matter and mind.

The pathological complexity thesis is simply an attempt to naturalize consciousness in all of its phenomenological complexity in a squarely Darwinian framework. To have phenomenological experience *requires* hedonic evaluation. To naturalize the puzzling notion of "qualia" is simply to explain how and why organisms have such a form of evaluation. Phenomenal states that do not appear to fit the evaluative dimension of consciousness can nevertheless be explained within the context of the evolution of Benthamite creatures. With evaluative capacities gaining in discriminatory sophistication that allows for the distinguishing and comparative evaluation of states, we can offer an explanation for why some sensory processes are conscious, or rather *felt*, whereas others are not.

There is thus no need to eliminate the notion of phenomenological experience, which Sytsma and Machery think is as flawed as the vitalist's concept of life (p. 322). Admittedly, there is often only a fine distinction between those who try to naturalize a concept and those who seek to eliminate and replace it. Perhaps the only difference is a degree of sympathy shared with the foregoing work of both philosophers and scientists who studied the mind and in particular our phenomenological experience. In thinking about the origins of mind in hedonic evaluation, however, the explanatory gap will no longer appear as large – making qualia less mysterious than they might at first appear. Inevitably, this will involve some reshaping of how our ordinary concepts of both "mind" and "matter" conceive of what goes on in organisms. But that is simply what it means to find the place of consciousness in nature. Some aspects of our folk concepts of these notions might be naturalized, whereas others are eliminated, ultimately providing us with a new scientifically informed view of the mind.

3.5 The Spoils of War

The goal of undertaking this war between the five dimensions was to discover and crown the core of consciousness. While we may lack direct paleobiological data that could indicate which dimension has the most direct line of descent to the dawn of consciousness, this chapter has attempted to reverse-engineer the origins of consciousness. Let me summarize which dimensions were lost and why.

I have argued that diachronic experience can be most readily dismissed as a necessary component of subjective experience since it is largely agreed upon to be absent in some animals, all the while granting them subjective experiences (of some kind). What Birch et al. (2020) called the temporality of experience can be seen as a higher-order feature of consciousness, not something that is likely to be present at its very origin.

Synchronic experience, while more often seen as a necessary component, must not be present at the very origins of consciousness either, since other animals without a strong connection between both hemispheres are likely to have a more disunified experience. There may be situations where disunity might in fact be more functional than highly centralized processing. Both dimensions of unity appear to be features of the way experience can be organized, rather than what makes them qualitative to begin with.

The experience of a self, I have argued, is something that is hard to disassociate from our thinking about consciousness, due to its centrality in our own conscious experience. But in a strongly gradualist picture, it is unlikely to have been present at the very origins, instead *built by* more basic kinds of subjective experience such as a distinction between exteroception and interoception within the dimension of sensory experience.

This left us with two contenders: the sensory and the evaluative sides of consciousness. Sensory experience has long dominated much of our thinking about subjective experience in philosophy and science. Yet, what this dimension seemingly fails to account for was the very *subject* that is so central to consciousness and makes the hard problem challenge seem especially damning to this dimension. It is then a natural scientific move to consider an alternative way of thinking in order to solve the problems of an old paradigm.

By drawing on recent work in experimental philosophy of mind, I have undermined the centrality of the sensory side in thinking about the most evident cases of experience. The *feelings side* which includes moods, emotions, and hedonic evaluations appears to be what drives the thinking of ordinary people about consciousness and thus provides additional support for the pathological complexity thesis I am trying to develop here. There is great untapped potential for building an alternative model of consciousness based on what are sometimes perceived to be the background features of ordinary human conscious experience: moods, pains, evaluations – features that usually only come into the centre when things go great or bad. This alternative way of thinking has unfortunately been largely resisted, yet has striking support from and has been partially developed across economics, animal welfare science, neuroscience, behavioural ecology, animal consciousness science, and ethology. Even if a theory based on the evaluative side of conscious experience will eventually turn out to be false, we are likely to make much greater progress by developing a picture that has received scant attention, yet ought to be considered as *at least* an equal competitor to the sensory-first views that have been modelled on the human phenomenon of visual experience. Unlike rich human-like vision, after all, a basic sense of evaluation – this primordial emotion – may have been present long before animals had any rich capacities to discriminate states of the world.

It is within an evaluative model of consciousness that it becomes straightforward to see how the other dimensions could be built on top of such a capacity. As Panksepp (2005) once argued: "affective experience may reflect a most primitive form of consciousness [...] which may have provided an evolutionary platform for the emergence of more complex layers of consciousness" (p. 32). These outer layers serve to enrich an evaluative mode of being that describes so much of animal life, with further gains in phenomenological complexity being united under a "common currency". And this is ultimately what motivates me to take a valence-first view of the evolution of consciousness: unlike the other dimensions, in an evaluative model of consciousness there is no explanatory leftover regarding how the other mysterious properties of consciousness arise. Whereas the strongly externalist and representationalist view of consciousness based on a model of visual capacities fails to account for selfhood and evaluation, the strongly internalist model of consciousness as self-awareness fails to account for the functional capacities of representation and evaluation. Both fail to put at centre stage the sensory-evaluative-motor feedback that is so central to a teleonomic Darwinian view of life. Such feedback ought to be at the centre of any theory of consciousness taken as a way of engaging the world as a vulnerable organism with a complex lifestyle that requires evaluation. Whereas defenders of strongly externalist and internalist approaches struggle to respond to advocates of the hard problem, that these capacities could just as well be executed unconsciously by an organism, evaluation fares better with this problem, since it is precisely the felt valence that makes such capacities functional.

Common-sense usage and introspection support the idea that hedonic valence is a fundamental feature of human experience, making it unsurprising that philosophers since antiquity have held similar views about the importance of pleasure and pain. And it is this evaluative dimension that ultimately won the war of the five dimensions. To unearth the evolution of the first evaluative sparks of experience by drawing on the pathological complexity thesis will be the target of the next chapter.

Notes

1 One excellent exception to this trend is Matthewson (2020) who argues that a trait may more or less have any single function, urging us to accept the graded nature of natural selection into our concepts of biological functions.
2 See Schukraft (2020).
3 See Tononi (2004, 2005, 2008, 2010, 2012a, 2012b), Balduzzi and Tononi (2008, 2009), Tononi et al. (2016), and Koch and Tononi (2011).
4 Since I won't repeat all their criticisms here, see Merker et al. (2021).
5 See Godfrey-Smith (2021a) for a discussion of these issues.
6 Tanaka et al. (2020), for instance, show that Lewontin's arguments for the role of the organism as a subject can be readily assimilated into more complex models of adaptive landscapes that include feedback between organism and environment.
7 See also Hurley (1998), Merker (2005), Godfrey-Smith (2016d, 2020b), and Trestman (2017).
8 See Lee (in press) for a recent critique of common arguments against a gradualist view of consciousness.
9 See Chalmers (1995) and Nagel (1974).
10 Though this is not to say that their case for two different concepts of consciousness in the public and among philosophers has gone unchallenged. See Huebner (2012), Talbot (2012), Peressini (2014), and Chalmers (2020) for challenges to their results, and a recent empirically supported defence of their view (Sytsma & Ozdemir, 2019).
11 I will explicate this idea in detail in Chapter 4.

4

PATHOLOGICAL COMPLEXITY AND THE DAWN OF SUBJECTIVITY

4.1 Introduction

The goal of the pathological complexity thesis is to explicate the evolutionary origins of "qualia" in evaluation. This idea, while intuitively attractive due to its direct link to fitness and action, is not as straightforward as suggestions by Darwinian thinkers such as Romanes may have made it seem. All life, as I noted in my response to LeDoux in Chapter 2, has organic states and processes that are beneficial or deleterious to it and some means of responding to these. Health is a universal biological measure of how an organism, whether a single-celled bacterium or a bat, succeeds at dealing with the species-specific pathological complexity challenges within their life histories. But this doesn't mean we should adopt the biopsychist stance and attribute sentience to all life, nor that we should take the stance of LeDoux and deny that such evaluative behaviour has anything to do with consciousness. Both sides fail to recognize a major transition in evaluative agency.

My argument is not that *any* degree of evaluative agency makes consciousness automatically appear, but rather that an explosion in pathological complexity through higher degrees of freedom comes to be dealt with through a major evolutionary transition towards a hedonic mode of evaluative agency, i.e. what I call *Benthamite creatures*. As Dennett once put it: "Complexity is worth caring about, but not just any complexity" (2017b). The complexity that is worth caring about is pathological complexity. Here, state-based behavioural and life-history theory offers us the ideal agential theory to emphasize the ecological importance of distinguishing biological success from failure, or in other words the adaptive from the pathological, which led to the evolution of consciousness as an adaptation for efficient action selection. Benthamite creatures evolve in the form of sentient beings with the real psychological utility to track this fundamental distinction between health and pathology, which makes the pathological complexity thesis a continuity thesis between properties of life and the mind. How and why this happened will be addressed in this chapter.

4.1.1 Chapter Outline

This chapter is organized into two main sections. In Section 4.2 "Complexity Worth Caring About", I address the objection of why it should be pathological complexity, rather than any other measure of biological complexity, that should matter for the evolution of phenomenological

DOI: 10.4324/9781003321729-4

complexity. In Section 4.3 "The Cambrian Explosion in Pathological Complexity", I seek to locate the origins of valence in the computational explosion of pathological complexity that occurred during the early Cambrian. Finally, Section 4.4, "Conclusion and Further Outlook" will summarize the conclusions of this chapter and sketch how the next chapter will explicate the pathological complexity thesis in more detail to naturalize the remaining dimensions of subjective experience within a model of valence as the most basic kind of "qualia".

4.2 Complexity Worth Caring About

At the beginning of this chapter, I have quoted a recent talk Dennett (2017b) gave at an Animal Consciousness conference at NYU. In this talk, he criticized the tendency among philosophers and scientists working on consciousness to only take the first step in explaining consciousness by analysing some of its properties, but leaving what he calls the "hard question" about what consciousness does *for* the organism entirely unaddressed. Agreeing with the widespread view that an understanding of pain and suffering would get us closer to understanding consciousness, Dennett nevertheless maintained that to understand the evolution of suffering, one has to ask *why* suffering matters. It cannot just be *intrinsically* bad, he said, maintaining that we must give an evolutionary explanation of why it matters to the organism. In order to explain the evolution of subjective experience, we must take a deep dive into the ancient origins of valence and examine what made a conscious mode of evaluation worth having. My answer to this question lies in an explosion of complexity that took place during the Cambrian explosion. But as Dennett emphasizes, we must answer which kind of complexity is worth caring about; it cannot be just any kind of complexity. That the answer to this teleonomic question is the normatively loaded sense of *pathological complexity* will be defended in this section.

In trying to reconstruct the possible evolutionary origins and function of consciousness, we are engaged in the paleobiologist's effort of making sense of a trait by connecting its extant "users" with its historical traces. When we think about valence in humans, it is choice, desires, motivation, and preferences that come to mind, and it is the evolution of such capacities, related to action, that we will have to pay attention to if we want to understand the dawn of subjectivity. One important figure – who has attempted to develop a plausible natural history of the evolution of human agency and cognitive states resembling the standard folk-psychological states of belief and desire – has been Sterelny (2003), but he unfortunately paid comparatively less attention to desire-like states, a lack that has also been criticized by David Spurrett (2015). Both Sterelny and Spurrett draw on the "ancestor" of the pathological complexity thesis to make sense of the evolution of preferences: Godfrey-Smith's (1996a) *environmental complexity thesis*. This was an attempt to make tenable within the modern framework of evolutionary theory earlier ideas from John Dewey and Herbert Spencer about the continuity between life and mind: the mind seen as a natural consequence of the evolution of living complexity.[1] Unlike Spencer and Dewey, however, who intended to include consciousness in their explanation of mental complexity in terms of living complexity, Godfrey-Smith restricted himself to explaining only basic cognitive capacities, excluding subjective experience. While the pathological complexity thesis differs both in its explanandum and explanans, it is indebted to the elegant explanatory framework and naturalist ambition of Godfrey-Smith's thesis. Furthermore, we may well ask whether environmental complexity could explain the evolution of valence as opposed to pathological complexity. After all, environmental complexity is routinely implicated in the evolution of desire-like states, behavioural flexibility, or, for that matter, organismal complexity.

However, while the environmental complexity thesis has an explicit link to action, it was (at least originally) designed as an externalist and representationalist theory. In that respect, it shared more with Spencer's externalism than it does with Dewey, who saw the complexity of the mind as something that evolved to deal with problems emerging in the dynamics between organism and environment (Godfrey-Smith, 1996a), which is closer to the pathological complexity thesis. Explications of the environmental complexity thesis have tended to pay very little attention to the organism as a "design and control architecture", instead treating the mind as something that decides what to do with the body, conditional on a given external state of the world (Spurrett, 2020, p. 5). While my book is largely concerned with consciousness, rather than cognition, the two are tightly linked, so it is perhaps unsurprising that I resist such a strongly externalist picture for cognition as well as consciousness. Godfrey-Smith's earlier motivation for a teleonomic theory of mind was similarly motivated by tying together externalism and adaptationism, which, while admittedly common, should not be accepted as a necessary marriage. A teleonomic view that tries to make sense of the mind within a theory of the organism will have to pay much more attention to the features *of* the organism.[2]

In making the emphasis on the organism clearer, it is useful to mention a largely inverse version of Godfrey-Smith's environmental complexity thesis that has been popularized by Keijzer (2015) and Keijzer et al. (2013). Whereas Godfrey-Smith's account of the origins of cognition largely idealizes away the organism, because of its inspiration from the externalist strategies of behavioural ecologists and evolutionary biologists, Keijzer's approach is inspired by the more internalist traditions of neuroscience and developmental biology. Keijzer et al. (2013) propose that the early function of the nervous system was to enable action "as a *single multicellular unit*" (p. 68). In this picture, sensing – and thus the input-output story – is relatively unimportant when we are looking at the very origins of the nervous system. The nervous system here largely plays the role of coordinating the body, irrespective of what goes on outside, going hand in hand with the evolution of contractile tissue (muscle) close to the skin or epithelium of an animal. Here, the nervous system evolves in order to solve a non-trivial control problem that has to be reinvented at a multicellular level. In a later paper, written as a direct response to Godfrey-Smith's earlier work on the environmental complexity thesis, Keijzer and Arnellos (2017) explicitly describe their hypothesis as an internal complexity proposal: "acquiring the fundamental sensorimotor features of the animal body may be better explained as a consequence of dealing with internal bodily – rather than environmental complexity" (p. 421). Despite the apparent conflict between these views, however, one need not necessarily see them as competitors. Spurrett (2020), for instance, argues that we could see them as two versions of a more general view, merely differing in their emphasis.

But what would a general version of such a view imply? *A* complexity thesis? This much, we already knew. Complexity, as Dennett noted, matters, but what kind of complexity? Spurrett (2020) proposes the "friendly amendment of the ECT" that takes into consideration both internal and external sources of complexity: "The function of cognition is to enable the agent to coordinate its (possibly complex) capacities, which can include coordinating those capacities with environmental complexity" (p. 5). But this definition only takes us half of the way. It's not enough to describe what cognition does on a general level. Like Dennett, we should ask the hard question: and then what happens? Why coordinate? Why act? It cannot be complexity *per se*.

In more recent reflections on his earlier work, Godfrey-Smith (2017b) admits that he was overly eager to state the environmental complexity thesis in externalist terms and recognizes that it isn't environmental complexity, *per se*, that matters, but rather the complexity faced *by* and mattering *to* an organism. Here, we ought to reject the dilemma between externalism and internalism. In the context of admitting his erroneous bet on externalism, the extensive discussion of Keijzer in Godfrey-Smith (2020b) may be seen as an attempt at self-correction and taking internalist views more seriously. He explicitly acknowledges Keijzer's influence in rejecting the mainstream representationalist thinking in the philosophy of mind, describing his ideas as an "emphasis on the *shaping of action*" (p. 59) that is so important if one wants to understand the branching of an animal *way of life*.

While Keijzer and his collaborators are a little too caught up with internalist ways of thinking, they are correct to locate the origins of mind in the control of action, rather than as Sterelny (2003) argues in the evolution of rich sensory detection systems. A Darwinian approach to the mind must ultimately emphasize dynamic feedback between organism and environment, since it is here that a kind of complexity dynamically emerges that *matters* and this can only be found in a teleonomic measure of complexity based on a theory of the organism, i.e. the *dynamic* life-history strategy of the organism. The problem we are faced with is simply the pathological complexity described in state-based behavioural and life-history theory. Features such as environmental and bodily complexity must be seen as variables that are relevant for the pathological complexity of an organism – they may be sources for it – but in asking for the complexity that matters to the organism, it is ultimately pathological complexity as *the* teleonomic measure of biological complexity that we must focus on. In asking what kind of complexity matters for biological agents such that the capacity for suffering is worth having, Dennett offers an elegant answer:

> The complexity of an autonomous, self-protecting, self-advancing (but mortal, vulnerable) bit of machinery gives us an explanation of why it is equipped to suffer, and why its suffering matters *to it*.
>
> *Daniel Dennett (2017b) [italics added for emphasis]*

It is only in a system that is (i) capable of "experiencing" biological damage from pathological states in the sense of "suffering" and (ii) capable of acting to avoid such states that it is worthwhile to be capable of *experience suffering* in the subjective senses of these terms. Inaction towards avoiding biological harms is itself pathological and gives rise to a new dimension of pathological complexity that can help us to make sense of why Benthamite creatures evolved. It is this teleonomic complexity that matters for organisms having to make decisions that lead to biological success or failure: how to avoid pathological behaviour and perform the right action at the right time? It is here that I shall argue that consciousness comes to perform its adaptive rationale.

4.2.1 Pathological Complexity and the Need for Valence

The pathological complexity thesis maintains that the function of consciousness is to enable the agent to respond to pathological complexity, which is the trade-off problem faced by all organisms as they pursue their goals. While pathological complexity constitutes a general design problem on the timescale of organismal evolution, the evolution of behaviour makes pathological complexity a constant problem for organisms to solve throughout their own lives, and it is within this explosion of pathological complexity in the evolution of multicellular action that I shall argue hedonic valence finds its origin.

Out of the many thinkers who have thought about the evolution of consciousness, Dawkins perhaps came the closest to articulating the kind of teleonomic complexity thesis that I defend here. Consider the following illustrative quote:

> Animals usually have more than one kind of danger to avoid. They have *complex tradeoffs* at all levels in order to minimize reductions of fitness in facing a wide range of threats. At different times of the day or year, or depending on external circumstances, they

> will reallocate priorities: For example, animals may depress or enhance their immune responses, increase or decrease their physiological "stress" responses, or find some stimuli more or less aversive.
>
> *Marian Dawkins (1998, p. 322) [italics added for emphasis]*

Like our focus here on evaluative feelings, Dawkins (1998) suggested early on that the very "key to the origin of consciousness itself may lie in the emotional experience of suffering" (p. 324). Notice that Dawkins speaks here of an experience *of* suffering – by which she means physical suffering in the sense of ill bodily health – rather than suffering *as* a mental experience, which can be interpreted as the suggestion that consciousness evolves first and foremost to respond to threats to health. Trained as an ethologist under Tinbergen at Oxford, Dawkins has been one of the most fervent critics of the lack of evolutionary thinking within animal welfare science. In an influential paper with the title "Evolution and Animal Welfare" in *The Quarterly Review of Biology*, Dawkins (1998) argued that to truly understand the welfare and subjective experience of animals, we must use an evolutionary approach as we would for any other biological phenomenon, "[a]nimal welfare, in other words, needs a dose of Darwinian medicine (Nesse & Williams, 1995)" (p. 305). Indeed, that there could be a strong evolutionary link between physical and mental suffering, or health and negatively valenced experience, has long been a central tenet of those working in the sphere of evolutionary medicine.

Nesse and Williams (1995), for instance, argued early on that some of the emotional states we disvalue and consider to be indicative of poor well-being – such as pain and fear – are evolutionary adaptations that are "unpleasant by design" (p. 26). Consciousness itself can be seen as just such an adaptation to ensure the health of the organism. Evaluative agency evolved in order to deal with the economic decision-making trade-offs of a new and flexible animal way of being. What Dawkins (1998) highlights is the need to recognize a teleonomic notion of the "complexity of an animal's adaptive response to various dangers" (p. 322), to which one should also add opportunities. Just as life is an evaluative and goal-directed activity, so is consciousness an evaluative and goal-directed way of engaging with the world, evolved within the context of life. It has evolved in order to respond to pathological complexity, which includes both opportunities *and* problems – such as the possibility for a common brushtail possum (*Trichosurus vulpecula*) to steal an unsupervised fledgling from a nest. What presents an opportunity for the possum also presents both a problem and a danger to the chick, and this ecological feedback between different organisms increases pathological complexity once more.

But we need to be careful not to become too tempted by the representationalist modes of thinking that turned the sensory side of subjective experience into the mainstream model for consciousness. Despite his earlier criticism of the environmental complexity thesis for tying adaptationism and externalism together,[3] Sterelny (2003) maintains that the environmental complexity thesis offers us something like a useful coarse-grained abstraction for investigating the origins of desire-like states. This is largely explained by his interest in the evolution of proto-representational states, and at least in this sense Sterelny is firmly connected to older mainstream representationalist thinking in the philosophy of mind. Spurrett (2015) likewise operates in a strongly representationalist model of the mind and considers the evolution of preferences and common currencies as value representations, though he at least recognizes that the problem of coordinating the body around action is very difficult and much neglected. Nevertheless, it is a mistake to tie the origins of valence together with the origins of representational richness, since it underestimates the importance of efficient decision-making and action control in the evolution of any system with higher degrees of freedom. In the absence of efficient action control, richer sensory capacities are simply a cost that is not worth paying.

Hedonic valence, as a commanding sensation, is plausibly much simpler than the cognitive processes we associate with sophisticated perception, possibly arising as something very primitive and immune to the demands of Lloyd Morgan's canon, but not so simple as to make it a default for all evaluative processes of life. By taking a design stance, i.e. by explicating the pathological complexity of different organisms and thinking about the properties that would make valence worth having, it is easy to see which properties would be relevant. Vulnerability and mortality *matter.* If a system is "indestructible" and almost immune to dangers posed by its environments, there is little sense in demanding pain, or for that matter pleasure. Autonomy and sufficiently flexible behaviour likewise *matter* because valence evolved to deal with the complexity of choice-problems. A system that cannot respond adequately to dangers or injuries, however, does not appear to require the machinery for evaluation. A capacity for negative valence has little to do with representing the world, not even internal states, and much more with enabling efficient adaptive behaviour that matters to the whole agent. The task of the pathological complexity thesis must be to turn this vague, but popular idea into a precise scientific hypothesis and framework for an ethological science of consciousness.

How organisms ought to deal with their species-specific pathological complexity can be explicated in terms of a unified teleonomic state-based and behavioural life-history theory of organisms that accounts for all the actions an organism can take. But Mangel and Clark (1986) rightly note that anything like a unified foraging theory will become almost impossible to assess, since "more complex models can rapidly become computationally unwieldy" (p. 1135). Typically, behavioural ecologists constrain the option-space of different actions organisms can take to a manageable set. But this is simply an idealization to make the life-history trade-offs manageable to model. Once organisms can take alternative actions that change their place in nature, we are faced with a dynamic programming problem, and "[i]t is well known that dynamic programming problems become computationally infeasible as their dimension increases" (Mangel & Clark, 1986, p. 1128). We are faced with a combinatorial explosion in trying to model to optimal life-history strategies for such organisms with high degrees of freedom. But the very reason it is so hard to model the maximization problem of their life-history strategy is why valence evolved as a proximate common currency for action selection. Pathological complexity is an optimization problem for organism and modeller alike: it is the complexity that matters for the *organism* both in the sense of a subject and object of evolution.

Within biology, some types of dramatically fluctuating environments are thought to favour very simple organisms: if it is very hard to survive bad seasons, the best option may be to have a capacity to reproduce very quickly when times are good (Bonner, 1988, p. 49). Godfrey-Smith's environmental complexity thesis neglects this: one way to deal with environmental heterogeneity is to become simpler and reduce the *pathological complexity* an organism has to deal with. To idealize away important features of the internal complexity of organisms in the context of understanding the function of mind forces us to neglect some of the most important features of what makes the organism an agent. In order to understand these complex dynamics of a teleonomic system, a state-based approach is needed that pays close attention to the life cycles and goals of organisms.

If we follow the ethologists' demand to study adaptive value alongside mechanisms and developments, we must answer the black box problem of how organisms optimize their behaviour. How organisms ought to deal with trade-offs between different goals is in principle no different from how we think conscious agents ought to resolve their conflicting goals. As Okasha (2018) notes, the agency in folk psychology, economics, and evolutionary biology typically requires "unity of purpose" or at least consistency among goals. We can usefully describe a system as an agent if there is a goal that all the processes and mechanisms work towards. The goal of organisms is ultimately reproduction and the pathological complexity thesis can thus describe them as agents with fitness providing a common currency for evaluating the importance of their different needs. However, as Samuelson and Swinkels (2006) rightly argue, organisms cannot just represent their fitness-function to achieve their goal of reproductive success since the complexity and lack of informational transparency of their situation make it impossible "to make the agent a *perfect* information processor" (p. 139). Natural selection was constrained to come up with utility functions over actions in a variety of life situations that in many cases will not directly map onto fitness, but nevertheless function in very analogous ways. Edmund Rolls (1999) made a useful distinction here between the kinds of choice mechanisms that are more fixed and found in organisms like plants and those choice mechanisms that are sensitive to learning in the achievement of a goal. Nature is not transparent and the values of different actions have to be learned (and often *un*learned), which is why Dawkins (2001) thinks that Rolls' distinction can help us to better make sense of the role of consciousness in the choice problems faced by animals.

What both biopsychists and LeDoux (2019) fail to recognize here is the evolution of a major transition in agency: agency evolving as a natural phenomenon, rather than just a general property of all organisms with the goal of maximizing fitness.[4] To learn about the workings of internal mechanisms that achieve this end of teleonomic action selection, behavioural ecologists will need to recognize that we need something like a common currency of fitness to compare different actions:

> Any attempt to understand behavior in terms of the evolutionary advantage that it might confer has to find a "common currency" (McCleery, 1977; McFarland & Sibly, 1975) for comparing the costs and benefits of various alternative courses of action.
>
> *McNamara and Houston (1986, p. 358)*

As I previously mentioned, common currency claims are common in the behavioural sciences, and Rolls (1999) similarly maintained that "a common reward-based currency appears to be the fundamental solution that brains use in order to produce appropriate behaviour" (p. v). Furthermore, several scientists have linked this idea to consciousness, including Merker (2007), Ginsburg and Jablonka (2019), and Solms (2021). The suggestion by Panksepp (2005, 2011) to focus on what he called the SEEKING system as an evolutionarily old but conscious motivational pull involved in the foraging activities of animals can also be seen as such a view. Likewise, Humphrey (2011) has argued that the origins of consciousness might lie in the intrinsic valuation or "enjoyment" of continued survival, an idea that admittedly fits well with my notion of Benthamite creatures.

However, the view I defend here of minimal hedonic valence at the "dawn of qualia" owes the most to Michel Cabanac, who has unfortunately received comparatively little attention, despite having arguably done the most to link the idea of a common currency to consciousness. Indeed, he has perhaps been the most outspoken contemporary defender of the old utilitarian *Benthamite* idea that this proximate common currency is the hedonic experience of pleasure and pain. This, he argues, is implicated in the early evolution of sentience in the early Amniota, which are estimated to have split from the rest of life around 330 million years ago (Benton & Donoghue, 2007). Together with his collaborators, Cabanac has been one of the first to emphasize the importance of positive and negative feelings in decision-making trade-offs in both humans and non-human animals (Balasko & Cabanac, 1998a, 1998b; Cabanac, 1971, 1979, 1992, 1996, 1999; Cabanac et al., 2009; Cabanac & Johnson, 1983). This is a different kind of common currency claim from that of McNamara and Houston, which is not about the problem of how behavioural ecologists ought to model the economic problems faced by organisms, but rather of how organisms themselves deal with their economic trade-offs. Here, the common currency is a real psychological state:

> In natural settings, the goals competing for behavior are complex, multidimensional objects and outcomes. Yet, for orderly choice to be possible, the utility of all competing resources must be represented on a single, common dimension.
>
> *Shizgal and Conover (1996, pp. 37–38)*

While Shizgal and Conover do not argue that such a common currency must be conscious, one can readily see why Cabanac connects such common currency claims to argue that hedonic valence will be able to function as a mechanistic proxy that mirrors the structure of the fitness-maximization problem space, and this is why the work of Cabanac is perhaps the most important inspiration for my attempt to find the origins of consciousness in hedonic valence.

Such a proximate common currency adds functional value by making the complexity of the computational problem tangible, enabling organisms with a large behavioural option-space to weigh alternative courses of action against each other. Organisms are often faced with what microeconomics studies as a so-called substitution problem.[5] Some needs and motivations are substitutes, i.e. one can be satisfied (at least partially) by satisfying the other. Others, such as sleep and foraging, conflict and need to be evaluated against each other in terms of importance. Benefits of one action need to be computed against the costs of foregoing another and the difficulty of this pathological complexity is rarely given enough appreciation, as if accurate representations of the world alone could fuel adaptive success.

Both Godfrey-Smith and Sterelny – in their emphasis on environmental complexity in the evolution of mind – have neglected this "internal" source of teleonomic complexity and underestimated the difficulty of achieving efficient action selection (see also Spurrett, 2020). This is why the pathological complexity thesis treats phenomenological complexity as something that must ultimately be subservient to evaluation, since it is here that it becomes discharged in action. Like Cabanac, I maintain that pleasure and pain are central in the evolution of animal consciousness, though I will argue that the evolutionary origins of this capacity are quite a bit older than he suggests, since i) – as I have argued in the previous chapter – we should not confuse the very origins of valence with its later function as a common currency and ii) Cabanac's argument that the increased complexity arising from terrestrial environments for amniotes made this capacity worth having underestimates the pathological complexity of underwater life ever since the Cambrian.

4.3 The Cambrian Explosion in Pathological Complexity

> [I]t seems certain, as a matter of observable fact, that the association of Pleasure and Pain with organic states and processes which are respectively beneficial and deleterious to the organism, is the most important function of Consciousness in the scheme of Evolution. And for this reason I have placed the origin of Pleasures and Pains very low down in the scale of conscious life.
>
> *George John Romanes (1883, p. 111)*

If dealing with pathological complexity is the raison d'être of valence, it is tempting to examine the history of life for explosions in complexity that plausibly made this capacity worth having. Doing so changes consciousness from being just a problem biologists may or may not want to address as an explanandum to an explanation for a rise in biological complexity itself. The most rapid and puzzling explosion of complexity in the history of life is the Cambrian explosion 541 million years ago. The *Cambrian Explosion* is important for the evolution of animals, since it is here that we observe the origin of all the major animal phyla we see today (Maloof et al., 2010). Stephen J. Gould (1996) even argued that it constituted the highest degree of diversity in animal life forms, making it maximally *disparate*.[6] The label "Precambrian" emphasizes the importance placed on the Cambrian period, seemingly representing the evolutionary equivalent of the Christian practice of identifying all time prior to the alleged birth of Jesus as "BC".

What caused this explosion of complexity, however, remains contested. This lack of a satisfying explanation has led some scientists and philosophers to seriously consider the possibility that we may be able to feed two birds with one scone by suggesting subjectivity, agency, and other capacities related to consciousness as a (partial) explanation for the Cambrian explosion (see, for instance, Feinberg & Mallatt, 2016; Ginsburg & Jablonka, 2019; Godfrey-Smith, 2016a; Trestman, 2013). If they are right, this would be an early birth of consciousness, indeed. Over the "short" time span of the 20 million years that followed, the Cambrian explosion led to complex multicellular body plans, nervous systems, behavioural repertoires, and modes of sensing.

4.3.1 A New Mode of Being

What we see in the Cambrian is the emergence of a new animal lifestyle in which agency and subjectivity come to play a crucial role: a "different mode of being" (Godfrey-Smith, 2020b, p. 79).[7] It is these features that come to the mind of most people when they hear the word "animal": active, sensing, and mobile creatures. Indeed, some think that all such beings are sentient. The folk usage of the term "animal" as a reference to living multicellular entities capable of goal-directed movement is thus not completely misguided when it is used to refer to something like a different *mode of being*. Aristotle, who was not aware of the microbial world of life where these capacities can also be found, used the properties of motility, sensing, and goal-directedness to distinguish a special animal mode of being from those of plants, whose mode of being consists in self-maintenance, growth, and reproduction. He called this animal mode of being the "sensitive soul", to be distinguished from the merely "nutritive soul" of plants and the "rational soul" that humans possess in addition to the other two (Aristotle, 1991), which influenced Ginsburg and Jablonka (2019) to title their book *The Evolution of the Sensitive Soul*. This way of thinking about animals treats them as possessing something *extra*, rather than a mere branching in the tree of life. While LeDoux (2019) is right that even single-celled organisms exhibit goal-directed survival behaviour and sensing, he fails to recognize that there is a significant major transition in the evolution of a distinctive animal lifestyle, and it is here that I locate the origins of sentience similarly to Ginsburg and Jablonka (2019) or for that matter Godfrey-Smith (2020b).

From an evolutionary perspective, the metazoan branch of life is much older than the Cambrian, plausibly already branching off from the rest of life 800 million years ago in some more or less recognizable transition towards multicellular individuality.[8] During the Ediacaran, which began roughly 635 million years ago and ended with the Cambrian, we find the first definite animal fossils, but they are largely plant-like and their behavioural capacities were simple (Peterson et al., 2008). The ancestors of such lifestyles are, of course, still around us now. Godfrey-Smith (2020b) vividly describes how scuba divers will inevitably encounter something like a "breathing forest" when encountering a "garden" of sponges, corals, and anemones, which are located somewhere between a plant and animal *lifestyle*, yet do belong to the animal branch of life.

In the evolutionary scenarios advocated by the aforementioned defenders of an early view of the dawn of consciousness, much of the focus has been placed on interactions *with others*. This makes a lot of sense in a sensory-focused view, since it is here that we see the evolution of sophisticated eyes and "tools", such as claws, for the engagement with other organisms. Interaction between *subjects* starts to *matter*; movement can become both "flight" and "attack". In their response to the first introduction of the pathological complexity thesis (see Veit, 2022b), Merker et al. (2022) interpret pathological complexity as just this emergence of interaction and co-evolutionary arms races in the Cambrian:

> Veit proposes that consciousness arose as a means for organisms to deal with what he calls pathological complexity. We assume that what he has in mind is the kind of complexity that arises in coevolution and evolutionary arms races, say of the predator–prey kind, which became acute with the evolution of large, image-forming eyes, hence his reference to the Cambrian Explosion.
>
> *Merker et al. (2022, p. 55)*

While this new dimension of interaction certainly leads to another explosion in pathological complexity by making the life histories of organisms vastly more complex, it is here that I locate the evolution of sensory experience, rather than the very origins of consciousness. As I shall argue, avoiding the overemphasis on interaction and sensing is a key distinguishing factor between my approach and competing views regarding the evolution of consciousness during the Cambrian. Taking a look at these competitors will help illuminate this difference.

Ginsburg and Jablonka (2019) see the Cambrian explosion as the driver of what they call *unlimited associative learning* (UAL): a special form of associative learning with a vast openness for new complex behaviour that they consider a transition marker for the presence of consciousness. This is because they think that UAL ties together eight features of consciousness widely acknowledged among philosophers and neuroscientists: 1) global accessibility and broadcast, 2) binding/unification and differentiation, 3) selective attention and exclusion, 4) intentionality, 5) integration of information over time, 6) an evaluative system, 7) agency and embodiment, and 8) registration of a self/other distinction (Birch, Ginsburg, & Jablonka, 2020, pp. 55–56). The details of this list do not matter much for the goals of this book and while I readily acknowledge that these features are important for the *shape* of consciousness, like Godfrey-Smith (2021b) I am not convinced of the idea that these features all need to appear together for consciousness to emerge.[9] While Ginsburg and Jablonka (2019) try to find minimal hallmarks of consciousness by excluding features unique to humans such as language, or at least only found in rudimentary forms in other animals such as meta-cognitive representations, we are still faced with the worry that the remaining features could exhibit a human-centric bias. As my discussion in Chapters 2 and 3 hopefully made clear, in order to avoid the idea that we must explain consciousness in terms of a certain rich human form of experience, for which all properties (whatever they are) must be there, or otherwise take the organism to lack subjective experience, we are in need of a comparative bottom-up approach that challenges combination views. The diversity of life should be reflected in very different ways of experiencing the world, so I am sceptical of putting too much emphasis on a certain combination of features that may be arranged in very different ways. UAL may well constitute a marker for an evolutionary transition of consciousness towards becoming more recognizably human-like, but the very basis of consciousness is more plausibly found in one of its properties, rather than the combination of a variety of capacities that have transformed consciousness across evolutionary time. For this, I emphasize what they list as their sixth hallmark: an evaluative system. The evaluative ability to avoid harmful stimuli and seek out beneficial ones is supremely important: survival matters. And some basic capacity for an imperative plus or minus "feel" can readily play important adaptive roles prior to any form of a combination of the above-mentioned "hallmarks".[10]

Others, such as Feinberg and Mallatt (2016), also offer an account of the origins of consciousness in the Cambrian, though their emphasis is on the evolution of eyes and exteroceptive consciousness as the original source of consciousness, which makes sense if one locates the origins of consciousness in the sensory dimensions. These approaches have so far failed, as Merker et al. (2022) rightly note in response to the pathological complexity thesis, to address the challenge of why "conscious vision, rather than simply better visually based performance operating unconsciously, is needed to meet the transition's functional challenge" (p. 55). My answer is that the origins of consciousness lie in the functional role of dealing with the complex trade-offs arising from the earlier explosion in pathological complexity due to the demands of controlling a multicellular animal body.

Lastly, as I noted before, Godfrey-Smith (2020b) primarily emphasizes agency and subjectivity, but these do not constitute a single property. They constitute a variety of capacities that can be described as making organisms more agent- or subject-like in some respects. While a detailed evolutionary journey from more object-like organisms to genuine conscious subjects will inevitably involve the gradual evolution of subjective experience and make consciousness less mysterious, it tells us little about the raison d'être and very origins of consciousness, unless we investigate the evolutionary origins of capacities that make organisms more subject-like.

While I think that all of the above approaches are important for understanding the evolution of phenomenological complexity, my problem with all these theoretical proposals for the evolutionary origins of consciousness is that they take a starting point at which animals have already gained a degree of complexity that I argue is indicative of a transition of consciousness towards more complex representational capacities. They all already assume some basic capacity for action and sensing as a given, which they then argue leads to interaction driving an arms race in which subjective experience makes sense. But in all this focus on *inter*action, it is lost that action itself constitutes a major problem (as the work by Keijzer and colleagues nicely demonstrates). And it is the solution to this problem that caused the Cambrian explosion.

4.3.2 Action!

Some clarificatory remarks on my usage of the term "action" will be useful here, since it is in the evolution of action that pathological complexity explodes and evaluative experience arises. Following Spurrett (2020) and the ethologists, I treat "action" here in the teleonomic sense of any kind of functional activity biological agents produce in the usage of their degrees of freedom, rather than (as is common in much of philosophy of action and mind literature) as an exclusive term for intentional behaviour.[11] This view is deliberately broad, to capture boundary cases and include minimal senses of action, such as can be seen in plants producing chemical defences. We are sticking closer to work in robotics, artificial intelligence, and cybernetics, where the computational complexity of building a teleonomic system is readily recognized. After all, computers are faced with their own pathological complexity in terms of optimal design and the teleonomic obstacles to the achievement of their ends. The usage of biological terms such as "viruses" and "malfunction" is not at all problematic in this context. The difference lies merely in the goals of organisms versus those of man-made machines. Indeed, at the end of this book, I will offer some considerations for how the work here may bear on the questions of whether and how we could build sentient machines. This broader notion of action will help us to better think about the evolution of agency in the animal branch of life. To reiterate my previous slogan with this new vocabulary: the expansion of the organismal option-space – or as a cyberneticist might describe it the organism's degrees of freedom – is what causes a computational explosion in pathological complexity for modeller and organism alike.

Action, of course, was not invented by animals. Evaluations are found in single-celled bacteria that swim, sense, hunt, and make decisions in the broad teleonomic sense of action I employ here. Pathological complexity is a property of all life and is not restricted to animals. Yet, I do not follow the biopsychist path of using evidence for evaluations in bacteria as evidence for consciousness. Such thinking seems motivated once again by resistance to evolutionary thinking about consciousness as something that gradually "emerges". It is a "category mistake", as Ginsburg and Jablonka (2019) put it, to identify consciousness with cognition in all living systems (p. 460). But we should also not commit the mistake of LeDoux that the existence of evaluation in bacteria implies that consciousness could not have evolved to aid organisms in evaluation. Consciousness should be seen as an adaptive design solution to a computational explosion in pathological complexity that the animal branch of life was faced with, not something that exists even at the most minimal kinds of that complexity, or only at a level of human complexity.

As I've previously alluded to, Godfrey-Smith's *Metazoa* shows a notable shift towards taking the bodily challenge by Keijzer more seriously. Here, he describes action as having to be reinvented at a larger scale with new forms of coordination (2020b, p. 53). What LeDoux failed to recognize is that when evolution has to reinvent or discover something at a new level of biological organization, things can take a very different shape and become vastly more complex. The challenge of organizing a multicellular unit is vastly more difficult than the challenge of organizing a single cell, even if the outcome appears superficially similar. But once this challenge has been mastered at a new level of organization, a vast possibility space is being opened for new ways of life. Multicellular action "involves coordination across vast scales from a cell's point of view" (Godfrey-Smith, 2020b, p. 53). The origins of this invention in the animal tree of life can be seen in the ancestors of modern *Alcyonacea*, or soft corals, who largely stuck with *minimal action* in the form of a grasping behaviour. Here, I should reply to the same kind of question Godfrey-Smith responds to when he asks why we should emphasize actions of movement over the production of chemicals or other basic activities of life. He argues that controlled motion was a new landmark in innovation, a sort of major transition of action, that made the organisms objects of a new kind (Godfrey-Smith, 2020b, p. 55). We can make this more precise by defining this kind of transition as the control of an organism's degrees of freedom in the service of a functional end, e.g. feeding or moving in one direction over another. Rather than conducting multiple independent actions at the same time, this major transition of agency allows for whole-body actions to the exclusion of others, thus leading to the evolution of choice.

In a materialist picture, it is tempting to ask whether the function of the first nervous systems may reveal the origins and function of mind. The origins of the nervous system constitute a major transition in evolution, with an entirely new mode of existence on a multicellular level. A problem Keijzer and Godfrey-Smith have identified in a joint paper is the narrow consideration of few roles or functions the early nervous system could have played, with the "neural control of development and physiology" often being "sidelined or omitted in favour of an exclusive focus on behaviour" (Jékely et al., 2015, p. 1). One should note here, however, that bodily control is often understood in too narrow a fashion, as the mere linking of behaviour with the right environment, but that would be a mistake. The biological world has few hard boundaries and we see something of a gradual transition from nervous systems playing the role of internal organization to taking a more outward-oriented role, such that action slowly emerges out of development (Godfrey-Smith, 2002). This is why Spurrett (2020) notes that some activities of plants do constitute genuine behaviour, e.g. in a Venus flytrap (p. 7). But in a gradualist picture from discriminating development towards the active agency, it would be a mistake to follow the move of some plant scientists such as Gagliano (2017, 2018) and fail to recognize that a significant transition towards agency took place in animal life. I acknowledge the force of their arguments that the striking cognitive and behavioural capacities of plants have been given too little attention,[12] but the right move here is to strongly endorse a previously neglected gradualism, rather than deny important differences, which is a similar fallacy to that made by those who tie the presence of agency in all of life to sentience in a biopsychist manner. As the work of Keijzer emphasizes, two important innovations that largely came together in the transition to a distinctive kind of animal agency are the nervous system and muscles that tied animal bodies together in new ways and allowed for a new set of adaptive capacities to be built on top.

An evolutionary view of consciousness must recognize gradations, and this applies to both the origins of subjective experience and the origins of the nervous system alike. Since subjective experience is often closely associated with the nervous system, it is useful to think about the very origins of nerves and neurons. Here, the role of muscles has been underestimated, with nerve nets largely playing the role of controlling muscles in the service of adaptive behaviour (Keijzer, 2015). In such a picture, it makes sense for a rapid explosion of innovation to occur in the form of sensory organs and tools such as claws to engage with other organisms, but the actual explosion in complexity was enabled by a transition in the organization of action. It is here that we find the dawn of subjectivity prior to the dimensions of sensory experience and selfhood.

4.3.3 The Dawn of Consciousness

To demonstrate the significance and difficulty of this major transition towards a distinctive kind of animal agency, one has to look no further than the time span it took from the origins of animal life to a distinctive animal lifestyle. The first definite fossils of animals date back to the late Ediacaran, though it has been contested whether to even call them animals, with their resemblance to odd flower-like shapes, long before the evolution of plants. *Dickinsonia*, which has been one of the paradigm animals of the Ediacaran, does not appear to have eyes or appendages that could give rise to interesting new ways of sensory-motor couplings. Godfrey-Smith (2020b) describes the biological imagination of this puzzling period as quiet and placid, with no evidence for interaction: "There are almost no signs of predation – no half-eaten individuals, no sign of the built-in weapons, offensive and defensive, that animals tend to have now" (p. 64). More importantly, however, is the striking absence of action in Spurrett's (2020) sense of degrees of freedom with alternative uses. Genuine action selection does not appear until much later.

Some change to this actionless picture began in the late Ediacaran (575 million years ago) until the beginning of the Cambrian, with a discernible transition taking place in animals. Waggoner (2003) distinguished three periods: Avalon, White Sea, and Nama. Strictly speaking, these three names were used to denote three major "assemblages", i.e. findings of a collection of species that fossilized around the same time. But despite new data coming in, the paleontological picture of three distinct periods has largely remained. The first of these periods is the most important, since it is here that recent discussions in the field have placed a possible second explosion in animal complexity. Following comprehensive quantitative data analysis of the fossil evidence, Shen et al. (2008) argued that there was an "Avalon Explosion" in the Ediacaran morphospace, mirroring the Cambrian explosion. However, while the White Sea showed definite signs of bilaterian bodies and more discernible actions of crawling on the seafloor, the Nama largely sees the disappearance of these larger complex and mobile animals, before they returned with a vengeance during the Cambrian. Here, it is useful to ask for possible mechanistic explanations of why one explosion failed, whereas the other succeeded.

Why such animal lifestyles failed despite gradual increases in sensory-motor capacities has puzzled Godfrey-Smith (2020b), who seeks to ground consciousness in the gradual evolution of just such capacities. Shen et al. (2008) ask but do not answer the question of what "constrained the Ediacara morphospace from further expansion or shift in the subsequent White Sea and Nama assemblages?" (p. 84). While it is unlikely that this Ediacaran extinction only had a single cause, the pathological complexity thesis offers us an elegant though admittedly speculative explanation for why one explosion eventually failed, whereas the other succeeded. The answer is the necessity of an evaluating system, which enables the efficient deployment of the increase in behavioural complexity through the gradual increase in sensory-motor capacities. Whereas organisms in the White Sea failed to deal with the computational explosion of pathological complexity caused by the rapid expansion of their degrees of freedom and were thus confined to more stationary plant-like ways of life, the Cambrian saw the evolution of *Benthamite creatures* with hedonic valence serving as an impulse for efficient action selection at the level of the organism. While this hypothesis is certainly speculative, one can readily visualize how – once this problem of efficient action selection was solved – higher degrees of freedom became an opportunity rather than a problem. The design space for more complex behavioural capacities was opened up, leading to a rapid explosion in the Cambrian morphospace. Natural selection was free to "experiment" in this competition between Benthamite creatures.

Here, the status of my concept of pathological complexity as *the* explication of the teleonomic complexity of organisms once again becomes important. The complexity that matters for the organism is first and foremost a problem to be solved, not an adaptation in itself. The real problem – I hypothesize – that was solved during the Cambrian, but not the Avalon explosion, was an efficient way of dealing with the increase in complexity of action selection. Since natural selection can only act upon behaviour by modifying the architecture behind decision-making mechanisms, McNamara and Houston (2009) argued that we need to combine the mechanistic research of physiologists with the adaptationist research of evolutionary biologists into an integrated study of function and mechanism. In the mathematical framework of state-based behavioural and life-history theory, it is obvious that an increase of variables through higher degrees of freedom will lead to a computational explosion in the complexity of finding the best strategies. How can organisms solve this? This problem has been given far too little attention, despite the fact that more *agential* organisms have to solve a problem for themselves that natural selection usually solves "for" other types of life, i.e. how to engage in the *right* fitness-enhancing activities. For much of life, these are a given, but for animals with behavioural flexibility, there is a constant need to compare the returns and costs of various actions, opportunities, and dangers associated with both internal and external changes. The reason that I suspect the Avalon explosion "failed" is because these organisms did not come up with a design solution to pay off for this complex investment into behavioural flexibility.

The animals that came closest to the animal-like lifestyles that we see today (such as *Dickinsonia* and *Spriggina*) were likely to have been some sort of grazing animals with few degrees of freedom that moved slowly across a surface abundant in nutrients. As this resource ran out (over millions of years), these animals were no longer able to sustain their way of life (Godfrey-Smith, 2020b, p. 68). Thinking about Cambrian animals, it would have seemed sensible for these animals to turn to each other as sources of food, i.e. to evolve the kind of interaction characteristic of the Cambrian explosion. Yet, for this to happen, an efficient way of organizing degrees of freedom was required, which these animals simply lacked. They did not have a valence system with which to make behavioural complexity manageable within a decision-making bottleneck that resembles that of natural selection both for propagules and species. As Charles Sherrington (1906) argued early on in his work on the goal-directedness of the nervous system, organisms require some form of informational bottlenecking – what he called a *final common path* – in order to deal with the problem of coordinating competing actions. It is precisely here that hedonic valence makes sense as an efficient trade-off mechanism to enable adaptive behaviour (see also Cabanac et al., 2009). This in turn enabled the evolution of richer behavioural and sensory capacities, with valence serving as a common currency for evaluation. As McFarland and Sibly once nicely put it:

> [D]ecisions among different courses of action must be made in terms of a common currency, and weighted among a common set of criteria. The necessity for comparing the merits of different courses of action implies that there must be some "trade-off" mechanism built into the motivational control system. Since the trade-off process must take into account all relevant motivational variables, it is clear that the mechanism responsible must be located at a point of convergence in the motivational organization.
>
> *McFarland and Sibly (1975, p. 290)*

Now, it is probably too much to demand that everything we've called "action" goes through a single bottleneck; that kind of thinking takes us back to an older Cartesian materialist model of the mind – one with a homunculus and a Cartesian theater that Dennett sought to dispel. But Spurrett (2020) is right to insist that there is something of an intermediate position here, "a useful corrective to the tradition [...] that regards almost any convergence in a control system as a symptom of allegiance to muddled models of intelligence and cognition" (p. 11). Spurrett is referring here to the likes of Brooks (1991) who argued that representational higher-order processing for actions would lead to significant bottlenecks with delays or even paralysis and that for this reason the world itself can serve as its own best representation (see also Clark, 1997).

However, some bottlenecking is required for an animal lifestyle, since information about both internal and external states is at least to some extent opaque, and the execution of one action over another requires the combination of a variety of capacities that, as Spurrett (2020) rightly notes, itself includes trade-offs "between other possible allocations of individual capacities and combinations of them, over and above whatever the metabolic and other direct costs of this or that action might be" (p. 11). But this increase in complexity as a problem that has to be dealt with has unfortunately not received much attention.

In debates on the evolution of consciousness even in the Cambrian, action production is largely taken for granted (e.g. in Ginsburg and Jablonka, Godfrey-Smith, Feinberg and Mallatt), but it cannot be disassociated from action selection. In dealing with this major transition in teleonomic complexity, a new level emerges in the Darwinian hierarchy of levels, with actions becoming distinguishable into adaptive and pathological behaviour. In order to deal with this new Darwinian challenge, organismal decisions will necessarily be made (or rather filtered/narrowed) through a number of different sub-agencies for computational reasons, but much work in neuroeconomics strongly supports the idea that there is in fact something like a global common currency for a huge variety of choice types if not, indeed, all (Spurrett, 2020).[13] It is thus not surprising that Shizgal and Conover (1996) maintained that orderly choice or – as economists would put it – revealed preferences are indicative that there must be some form of value ranking on a common scale. And valence plausibly constitutes an ancient solution to this problem in the Cambrian, which then enabled the evolution of richer kinds of felt sensory representations, such as Denton's (2006) primordial emotions like thirst and hunger directly tied to an evaluative system of efficient decision-making. An evolutionary perspective turns on its head the common view that affect is something that came on top of sensory consciousness, by making hedonic valence its most ancient capacity. The pathological complexity thesis does therefore not just imply a metaphorical sense of the existence of a common currency, but a psychologically real *felt* common currency. And it is because of this that my account offers an elegant answer to the challenge of why some sensory processes are felt and others aren't.

Unfortunately, comparative neuroeconomics remains an incredibly small field, with much of its research focused on standard model organisms such as monkeys, rodents, and birds. But what little research has been done strongly supports the idea that analogues to the human final common path are found across a wide range of the animal branch of life. This is precisely where we'd expect the presence of a common currency, in the form of specialized neural circuits designed for the exclusion of mechanically incompatible actions that can compete against each other in a preference ordering and also, as Spurrett (2020) emphasizes, "the last place [preferences] can do so" (p. 22). It is now clear that almost all vertebrates share an evaluative neural system for reward and "punishment", with dopamine and other valence-related molecules such as serotonin sharing a deep evolutionary origin that is plausibly very ancient, rather than having been invented multiple times. The observation that some species taken to lack sentience, such as nematodes, use dopamine (alongside other such molecules implied in valance) to organize action and motor activity (see Barron et al., 2010) is only further evidence of an ancient origin of valence to help organisms achieve efficient action selection. As Spurrett (2020) notes, it is highly likely that the first implementations of preferences "were elaborations of motor control systems shared with creatures that couldn't learn, but could move" (p. 23). Indeed, it appears that dopamine is at least as old – if not older – than the invention of the bilaterian body plan with symmetric halves, which enabled a vast expansion of possibilities for animal action (see Caveney et al., 2006).

In the theory defended here, we might thus see a valence system as the revolution needed to make animal agency "pay off", providing an efficient action selection mechanism as the final behavioural common path in metazoans for the prioritization of some actions over others as the complexity increases through more degrees of freedom. What evidence has been gathered in invertebrates on evaluation and motivational trade-offs is highly suggestive that this is not a unique vertebrate trait. If not for the standard methodological practice in comparative cognition to only attribute these capacities to animals for which these capacities have been explicitly demonstrated, they would probably already be seen as much more basic. We are unfortunately here faced with something of a methodological artefact, in the observation that studies of motivational trade-offs (such as those of Cabanac and Elwood) have until recently been very rare in animals distantly related to us, though I will discuss promising recent research on bees in the next chapter.

Unlike Spurrett, who tries to bracket away consciousness and emotions in his work on the origins of preferences as representational capacities, I see them as being initially instantiated through a hedonic valence system that later became more representational by acquiring the richer sensory and integrative capacities discussed in the previous chapter on the dimensions of unity and sensory experience. These dimensions of consciousness are evolutionary add-ons, plausibly appearing in roughly the reverse order in which I have shelved them away. Indeed, it is this feature of a valence-first view that allows it to stand up to the challenge of Lloyd Morgan's canon, as it is something substantially simpler than the idea of meta-cognitive representationalist value rankings. Cognitive complexity or "intelligence" should not be seen as a prerequisite for simpler affective capacities (see also Browning, 2019). As Dawkins elegantly puts it:

> [Y]ou don't need to be very clever to feel pain or hunger or fear. Negative emotions that we refer to as suffering are not particularly intellectual. Indeed, we often have difficulty

> trying to subdue our emotions with our more rational thoughts. Is it therefore not possible that the evolutionary origins of consciousness are far older than those of more refined cognitive abilities?
>
> *Marian Dawkins (2001, pp. S27–28)*

Affective decision-making is plausibly more ancient than the cognitively-demanding sensory-motor couplings that evolved in the course of the evolutionary arms races of interaction. However, that doesn't mean that there isn't a design challenge here. After all, even if animals have something like subjective utilities, that doesn't mean that they will correspond to the fitness of the organism, even if this link has often been asserted (see Okasha, 2018). Establishing this link is not an easy task, as Rolls (1999) elegantly points out:

> Evolution must set the magnitudes of each of the different reward systems so that each will be chosen for action in such a way as to maximize overall fitness. Food reward must be chosen as the aim for action if some nutrient depletion is present, but water reward as a target for action must be selected if current water depletion poses a greater threat to fitness than does the current degree of food depletion. This indicates that for a competitive selection process for rewards, each reward must be carefully calibrated in evolution to have the right common currency in the selection process.
>
> *Edmund T. Rolls (1999, p. 267)*

Rolls and Damasio maintain that there are good reasons for thinking that these values will be updated through experience via reinforcement learning. Yet, we could question this idea. Is it really plausible that reinforcement learning can solve the complex state-based behavioural and life-history calculations that organisms are faced with? To a first approximation, we should reply with a confident yes. There is no need for subjective values to perfectly match the biological values an organism is faced with, as long as there is a sufficient degree of correspondence that enables organisms with hedonic valence to outcompete those without such a form of evaluation. Risk, in the economic sense of uncertain value returns from actions, is an inherent factor in biological life. Environments can change and cause what is often described in evolutionary medicine literature as an evolutionary mismatch (Manus, 2018). If the environment changes, an action that might previously have been fitness-advancing could now be deleterious in nature. In the early Cambrian, this factor would have only been even more important.

One way of establishing a link between valence and fitness is reinforcement learning. If something like a minimal capacity for hedonic valence comes into existence, there will be an almost immediate fitness benefit to linking this capacity to reinforcement learning. This would then also require that the subjective values can be updated, requiring further gains on the sensory side of experience. Evaluative experience would quickly become enriched in order to allow more fine-grained distinctions between what is good and what is bad for the organism, thus likely bringing the sensory dimension of consciousness soon into existence after the evaluative core has been established.

Dennett (1995a) once called organisms capable of reinforcement learning *Skinnerian creatures*, but a better term – one that is less reliant on externalist modes of thinking about what happened in the transition to a distinct kind of animal agency – would be *Benthamite creatures*. These creatures were moved by some kind of "quasi-intrinsically" motivating states of hedonic valence and thus contained the origins of qualia. Capacities for reinforcement learning are highly suggestive of preferences, implying "both sensitivity to rewards and updating behavioural dispositions in light of reward-based consequences of earlier behaviour" (Spurrett, 2020, p. 23). It is thus hardly surprising that earlier evolutionists took such learning abilities as convincing evidence that these animals can feel pleasure and pain. Natural selection plausibly led to an evolved imperative "tendency to repeat certain actions because they feel "good" (are positively reinforcing) and this feeling good or pleasure then guides the subsequent behaviour" of the agent (Dawkins, 2001, p. S23).[14] But because of the apparent ubiquity of this ability in the animal branch of life including cephalopods, crustaceans, and insects (Perry et al., 2013),[15] many have come to endorse the view that this would make consciousness too simple; that it requires something more.

But why should we hold this position? A multicellular organism that is reliably able to produce adaptive behaviour through reinforcement learning is not at all just a simple machine. The problem, as I see it, is rather how subjective evaluations can be mapped onto the actual fitness values prior to such an updating capacity evolving in Cambrian organisms. The answer, however, is not to think that Cambrian species were equipped with fixed and uniform utility functions over their sets of actions, but rather that there was great diversity even within a species that allowed species to faster exhaust the design space of hitting upon adaptive actions. This is why the Cambrian explosion showed such a fast expansion in diversity: there were immediate benefits to hitting upon more efficient means of action selection. Natural selection simply filtered out those individuals whose subjective evaluations had a worse fit to their "hidden" fitness values. A new realm of pathology was born, where minds could be described as pathological in virtue of how well (or poorly) they establish continuity with the biological reality of the organism – which is why I say that the pathological complexity thesis is a life-mind continuity thesis. New forms of learning quickly arise in this context to create a close fit between mind and world and to ensure the organisms' Darwinian success, but I don't see a very basic capacity of hedonic valence as something that at all requires complex capacities. Rather, it is a major transition enabler upon which new capacities can be tested, to make use of high degrees of freedom in the service of an organism's ends. It is the beginning of the organism itself becoming engaged in doing life-history calculations within its own lifetime. At the very dawn of this phenomenon, a mere + and – feel may not have accomplished this too well, but it opened the pathway for richer forms of evaluative updating to evolve, which would solve this Darwinian challenge for organisms within their own decision-making.

The evidential burden to attribute consciousness appears to be raised as soon as we learn more about the complex capacities of other animals that we previously assumed to not be conscious. But that is a flawed approach. Firstly, a gradualist evolutionary perspective of the evolution of consciousness ought precisely to endorse a very humble origin for the origin of sentience, and secondly, the ability for affective decision-making and learning is far more complex than is typically given credit, with cyberneticists struggling to design robots achieving even the most basic successes of simple animal life. No robot has of yet been created that would be able to handle the pathological complexity of the life histories exhibited in even the most basic of the distinctively animal lifestyles. Their failure is akin to the very same challenge Avalonian and White Sea animal agents failed to overcome.

What we see at the verge of the Cambrian explosion is the origin of Benthamite creatures with hedonic experience upon which more complex representational capacities, such as interoception, could be built. Here, I am not implying that all new capacities must go through the bottleneck of the evaluative system and be consciously experienced. But it is within the context of such evaluative agency that subjective experience *makes sense* and plays a distinctive role in the functional deployment of the degrees of freedom a flexible animal lifestyle offers. This approach substantially narrows the explanatory gap by breaking the phenomenological complexity of consciousness down to its minimal evaluative core. Instead of a strange, apparently unnecessary add-on, subjective experience is reconceived as something almost necessary to allow a major transition of multicellular organisms from "mere" objects subject to the whims of external forces into genuine agents/subjects and thus makes great progress in the completion of the Darwinian revolution.

4.4 Conclusion and Further Outlook

The goal of this chapter was to shine light on the close connection between consciousness and pathological complexity. The ethologists long emphasized that we need to understand organisms as teleonomic agents with life-history strategies in their natural environments. Without an understanding of what these organisms evolved *to do*, it will be impossible to distinguish the normal from the pathological. Yet, this is precisely the bottleneck through which evolutionary theorizing enables us to make progress on the scientific puzzles of life. From an evolutionary point of view, health has to be understood as a measure of how an organism deals with the pathological complexity it is faced with: it is the ultimate teleonomic measure of organismal complexity. And pathological complexity can be operationalized as the complexity of the number of parameters and constraints in the optimization problem studied by state-dependent or state-based behavioural and life-history theory. With the evolution of behavioural flexibility, the Cambrian explosion brought forth an explosion in pathological complexity of how to control action – a problem that I argued was dealt with through the evolution of hedonic valence. This in turn plausibly enabled the evolution of sensory consciousness as an enrichment in representation, providing the discriminatory richness of these Benthamite creatures, with valence serving the role of a proximate common currency. Similarly, self-consciousness plausibly evolved gradually out of further enrichments of these sensory capacities, distinguishing between self and other. Unlike strongly externalist and internalist views of consciousness, there does not appear to be some explanatory leftover – a feature of consciousness that we have left out. The explanatory gap has been significantly reduced by placing at the centre a dimension of consciousness that is inherently dynamic.

The strongly gradualist approach taken here should be attractive precisely because it minimizes the requirements for the most basic kind of consciousness, without making it an automatic feature of either life or a particular level of cognitive complexity. Consciousness, rather than simply appearing as a by-product of other capacities, gradually evolved as a special evaluative mode of being in the Cambrian; what I have called the evolution of Benthamite creatures. In principle, I admit, one could take the view that the "lights suddenly went on" in the evolution of this efficient way of responding to pathological complexity, yet what we see here is a major transition in animal agency with organisms gradually becoming recognizably more experiential.[16] The evolution of Benthamite creatures gradually turned "objects" into "subjects" with their own needs becoming *felt imperative interests*, thus bringing us closer to Lewontin's demand to bring the Darwinian revolution to completion by paying attention to the *functional needs* of organisms. The evolution of subjectivity is seen as an efficient solution to the problem of action selection in the control of a complex multicellular body, enabled through hedonic evaluation. This design solution unlocked a vast sphere of design space, including a distinctive subject-like animal way of life. Here, we see a new level of health and pathological complexity emerging beyond physiology and behaviour; the mental life of organisms can now itself be considered adaptive or pathological by virtue of how it promotes the fitness of the organism, thus giving rise to a new level of complexity, but also an opportunity for animals in their engagement with the world. The appearance of sentient Benthamite creatures gave rise to the diverse phenomenological complexity of electro-sensing platypuses, echolocating bats, infrared-sensing snakes, and the like that we find in nature today.

With this hedonic model in place, the next chapter will try to build a firm understanding of phenomenological complexity as a response to pathological complexity by placing the other dimensions in the context of a general theory of consciousness based on the evaluative side of experience. It will enable us to make predictions regarding the likely subjective mental states of animals based on the evolutionary history and ecological challenges faced by them *in nature*, which can then be tested through various experimental means.

Notes

1 See also Godfrey-Smith (1996b, 1996c, 1997).
2 Recall in Chapter 1 the criticism by Lorenz of the behaviourists' externalism that it leaves out everything that makes an organism an organism.
3 See Sterelny (1997).
4 This is also why I have criticized Okasha's (2018) monograph on agency as a concept in evolutionary biology, for having little to say on the actual evolution of agency as a real phenomenon in nature (Veit, 2021e).
5 See Shizgal's response to the question of whether there is a common currency for all sensory pleasures in Kringelbach and Berridge (2010, p. 18).
6 See Sterelny and Griffiths (1999) for a critical discussion of Gould's views.
7 See also Ginsburg and Jablonka (2019).
8 While I have previously written about the challenge of evolving multicellularity (Veit, 2019), the evolution of multicellular agency constitutes a distinct major transition of its own.
9 See also my essay review of their monograph for a more detailed analysis and critique of their account (Browning & Veit, 2021a).
10 See also Barlassina and Hayward (2019) and Barlassina (2020), who try to offer an imperative metaphysics of valence that, while I don't agree with all their arguments, nevertheless supports the picture offered here.
11 This parallels the use of "behaviour" as functional activity by Millikan (1995).
12 For a set of interesting findings about plants, see Gagliano et al. (2016), Gagliano (2017), and Calvo et al. (2020).
13 See Levy and Glimcher (2012, 2016) and Pearson et al. (2014) for excellent reviews of the neuroeconomics literature.
14 See also Rolls (1999).
15 Note that all these animal groups are now slowly entering the accepted realm of animals with sentience, and I will discuss their phenomenological complexity in the next chapter.
16 Godfrey-Smith (2020a) suggests the useful metaphor of "experientialization" (p. 215).

5

PATHOLOGICAL COMPLEXITY MEETS PHENOMENOLOGICAL COMPLEXITY

5.1 Introduction

We have reached the stage at which we can synthesize the previous chapters to make progress on the cognitive ethologists' goal of assessing the phenomenological complexity of other animals. With a model in place of consciousness grounded in the evaluative agency, it is time to return to the other four dimensions that we previously shelved away from the most ancient kind of subjective experience. It is here that we find perhaps the strongest virtue of the pathological complexity framework in being able to draw on modern life-history theory to make predictions regarding the subjective experience of other animals based on our ethological understanding of what it means to be an octopus, a bee, or raven in their natural healthy lifestyles. As Griffin once put it:

> Granting that hypotheses [about animal minds] are difficult to test by currently available procedures, the tentative consideration of their plausibility might pave the way for thoughtful ethologists to devise improved methods to study when and where animal consciousness may occur and what its content may be. The future extension and refinement of two-way communication between ethologists and the animals they study offer the prospect of developing in due course a truly experimental science of cognitive ethology.
>
> *Donald Redfield Griffin (1981, p. 171)*

But the goal of this chapter is not only to show that the other dimensions of consciousness can be readily explained within a theory of consciousness centred on evaluative experience. We will also put the pathological complexity thesis to the test by critically evaluating a suggestion by Godfrey-Smith (2020b, 2020c) that there could be a phylogenetic split in conscious experience, with some animals having an evaluative experience while lacking the sensory side or vice versa. In my continuous emphasis on phenomenological complexity, we should at least take seriously the "idea of deep differences between varieties of subjectivity" (Godfrey-Smith, 2020b, p. 217). If we find animals that only have sensory experience without the evaluative side, that would at least raise an interesting challenge to the pathological complexity thesis.

To test my thesis, we will first look at the gastropods (snails and slugs) and secondly the arthropods (in particular crustaceans and insects). Admittedly, only insects (hexapods) constitute a real test case, since Godfrey-Smith (2020c) uses gastropods as a potential case that "may have relevant evaluative

DOI: 10.4324/9781003321729-5

complexity" to be experienced but lack sufficient complexity on the sensory side (p. 1153). Indeed, my discussion of gastropods will primarily serve as evidence *for* the pathological complexity thesis: the possibility of minimal consciousness in the sense of hedonic valence, without the other dimensions of consciousness. In insects, however, Godfrey-Smith maintains that they have only simple evaluative capacities, whereas their sensory capacities are sufficiently rich to make it at least plausible that they could have sensory experience without the evaluative side. Reviewing the evidence in the literature for this view, I will ultimately reject this challenge to my thesis.

After this defence of the pathological complexity thesis, I will turn to the dimensions of unity and selfhood and discuss them in the context of the species-specific pathological complexity of cephalopods, non-avian reptiles, fish, and birds.

5.1.1 Chapter Outline

This chapter is structured as follows: in Section 5.2 "Gastropods: A Sluggish Way of Life", I use the case of gastropods to support the motivation of the pathological complexity thesis to seek the origins of consciousness in evaluation. In Section 5.3 "Arthropods: A Robotic Way of Life", I respond to the challenge that insects might have sensory experience without evaluative experience. In Section 5.4 "Octopuses: A Disembodied Way of Being", I discuss the dimension of self-consciousness through the biological lens of the octopus, who arguably provide the best biological challenge to human-centric ways of thinking about the role of the self. In Section 5.5 "Fishes and Non-Avian Reptiles: Dis-Unified Ways of Being", I discuss these natural split-brain patients to assess the possible adaptive benefits of more disunified forms of consciousness. In Section 5.6 "Corvids: A Cunning Way of Being", I discuss the adaptive roles of diachronic unity through a comparison of corvids and those humans with aphantasia. Finally, Section 5.7 "Challenges, Conclusion, and Further Directions" summarizes this chapter, offers some responses to potential objections, and explores potential directions for the further development of the pathological complexity framework.

5.2 Gastropods: A Sluggish Way of Life

The first class of animals I shall discuss are gastropods (i.e. snails and slugs), which form a large group of invertebrates and evolved in the sea during the Cambrian, though their precise origins remain contested (Parkhaev, 2007). Like the cephalopods, who are now accepted by many to possess subjective experience, gastropods belong to the phylum of molluscs, though their nervous system is generally simpler. Unlike the cephalopods – and in particular octopuses – gastropods have received only little attention in debates on animal consciousness. Despite their taxonomic diversity, their lifestyles have been considered too slow and too uninteresting, compared to the extreme behavioural flexibility, tempo, and intelligence of their octopus relatives. One might thus be tempted to categorize their pathological complexity as too insignificant. But here we should heed the warning by Dennett (2019a) that our imagination is in many ways shaped by what Wittgenstein dubbed *Lebensform* (translated: a form of life), i.e. "our linguistic communities, the commonalities that are apt to confound our thinking with parochiality" (p. 2). If we observe animals distantly related to us and with very different ways of life, we will be influenced by what Dennett (2019a) nicely expressed as their *speed* and *rhythm*:

> [I]f cephalopods moved in the clunky way of most existing robots, then in spite of the manifest purposiveness of their motions, it would be quite comfortable to suppose that they were some kind of zombies, marine robots with eight or ten appendages.
>
> *Daniel C. Dennett (2019a, p. 2) [emphasis in original]*

Gastropods, of course, appear even slower than many sophisticated robots. In this context, it is also worth pointing out that it is hard for many to avoid the illusory attribution of consciousness brought about by sophisticated ventriloquists[1] or robots designed to move in very human-like ways.[2] Care must be taken not to put too much faith in our intuitive willingness to grant consciousness to some entity. But one need not draw on thought experiments or artificial cases such as robots to get this point across. While Dennett (2019a) sees giant sea slugs of the genus *Aplysia* as nothing more but an excellent model organism due to their simple nervous system, Godfrey-Smith (2020b) replies from his personal experience that unlike other smaller slugs with similar nervous systems, "all one has to do is scale them up to giant *Aplysia* size and have them move at a gallop rather than a slow crawl, and suddenly experience in these animals seems almost inescapable, or at least far more feasible" (p. 216). What should one make of such observations? The lesson, I suggest, is that it is misguided to make our attribution of consciousness contingent upon our ability to occupy points of view of others. We are too influenced by our own distinctive human perspective.

Whether we are faced with echolocating bats, crawling gastropods, buzzing bees, or oddly moving octopuses, we have to remind ourselves that these animals live different *kinds* of lives, which ought to bear out in different kinds of minds. Their conscious experience should reflect their life histories, which are strikingly different from ours. Perhaps humans simply lack the imagination required for this task. We are, after all, notoriously bad at putting ourselves in the shoes of others – this not only applies to strangers of a different sex, race, or origin, but even to our closest family members. More notable perhaps is our inability to even recall what it was like to be us in the past or predict what it will be like to be us in the future. The best-known example of this is the *disability paradox*: while people with disabilities often report a high quality of life, non-disabled people believe this to be a cognitive mistake – imagining much bigger losses in quality of experience when they imagine themselves to have these disabilities (Albrecht & Devlieger, 1999). Introspection can be positively detrimental to the assessment of another being's subjective experience, even in our own case. Here, gastropods constitute a beautiful case for the need of an ecological bottom-up approach.

Prominently, Feinberg and Mallatt (2016) argue that evidence for consciousness in gastropods is lacking, but they also admit that there is some evidence pointing towards the affective side. Nevertheless, they end up denying consciousness to gastropods since they are said to "lack the brain complexity one would expect for consciousness" (p. 192). This, of course, raises the same validation problem we saw for *integrated information theory*, of whether we already know what complexity would actually be required for consciousness. Godfrey-Smith (2020c) evaluates the evidence in a different way by emphasizing that gastropods may be a case for sufficient richness in evaluative capacities to have evaluative consciousness while lacking the other dimensions. If so, this would provide strong support for the pathological complexity thesis: we could have animals around us in the here and now, rather than just at the origin of consciousness in the Cambrian, with a minimal sense of hedonic evaluation without the other dimensions. A theory of consciousness based on the human case is undoubtedly prone to fail in its recognition of such "marginal" cases, so it is useful to look at their "lived experience" from their own point of view by using the pathological complexity framework.

5.2.1 Evaluative Experience

In his emphasis on the evaluative capacities of gastropods, Godfrey-Smith draws particularly on the work of Terry Walters, who has been one of the front runners in advancing our understanding of gastropod *skills*.[3] As I've emphasized in my discussion of the evaluative dimension, we should not simplify this dimension to contain only pleasure and pain, in the sense of two highly specific mental states, but rather include any sort of subjective experience that has a positive or negative valence. This can also include medium-term and long-term states such as emotions of anger or fear and moods such as pessimism. As we will also see in the discussion of insects that will follow, we should be open to the existence of all kinds of negatively valenced states and not limit them to human-like cases of pain involving rich sensory representation.

Crook and Walters (2011), for instance, argue that gastropods show *nociceptive sensitization*, which Godfrey-Smith (2020c) describes as "a heightened sensitivity after damage" (p. 1155) and sees as compelling evidence for perhaps a minimal sense of evaluative experience. What this work has shown is that when gastropods are exposed to aversive stimuli such as electric shocks, they not only react to this with an immediate behavioural response, but there also appears to be a long-term change in behavioural "character". Crook and Walters (2011) argue that *Aplysia* shows a conditioned fear-like motivation state when exposed to neutral chemosensory stimuli when it has been associated in the past with an electric shock (p. 189). Furthermore, associative learning was demonstrated in gastropods as early as 1981 (see Carew et al., 1981; Colwill et al., 1988; Walters et al., 1981), an ability that has often been linked with consciousness. When the smell of a shrimp was paired with an electric shock, *Aplysia* showed surprising future responses to the stimuli, such as (i) freezing in response to the smell even in the absence of electric shocks, (ii) halting feeding when exposed to the smell, and (iii) withdrawal, escape, and defence responses when the smell was paired with light touch (Crook & Walters, 2011, p. 189). Godfrey-Smith (2020c) considers this range of responses compelling evidence for a "pervasive state of negative readiness" linked to the feelings side of subjective experience (p. 1155). From exchanges with Walters, Godfrey-Smith reports that he is more cautious about attributing sentience to them, but acknowledges the striking teleonomic rationale of an "ability to maintain functional "awareness" of injury-induced vulnerability until the vulnerability subsides (perhaps until adequate repair of damaged body parts has been achieved)" (Walters, 2018, p. 13) [cited in Godfrey-Smith (2020c, p. 1155)].

If the pathological complexity thesis is right, then this is exactly how the vulnerability of complex multicellular organisms gives rise to hedonic experience. One may even see these negative mood states as involving a minimal sense of self and diachronic unity, but these features need not be part and parcel of the subjective experience of an animal in order to associate particular stimuli with a negative valence. After all, even humans can have a negative emotional reaction to an event or food item without the ability to consciously draw the connection to a previous negative encounter. Nevertheless, it is tempting to think that episodic memory can be readily explained as something built on these capacities once they are in place, and we should resist the thought that current boundary cases for the attribution of sentience must be anything like the animals in the early evolution of subjective experience. It is not at all implausible to think that the presence of a hedonic evaluation system quickly gives rise to further increases in phenomenological complexity.

Furthermore, Godfrey-Smith (2020c) praises Walters for highlighting the ecological lifestyles of *Aplysia*, which often involve longer life cycles of one to two years more than is common in many insects. If we try to explicate the pathological complexity of gastropods, we will quickly find an additional rationale for these long-term mood states. Because their behaviour is relatively limited in comparison to that of many other animals that are discussed as potential bearers of sentience, wound-tending does not appear to be even within their behavioural option-space. Yet, this doesn't mean that gastropods aren't vulnerable. Unlike insects, whose bodies are hard, many gastropods lack even shells to protect themselves. But whereas insect bodies can often not be "repaired", hence making protection superfluous, gastropods almost constitute an opposite case, with excellent if not extreme abilities to heal. As long as wounds are not mortal, they will quickly restore their bodies to a healthy state of normalcy.

An extraordinary case in the genus *Elysia* cf. *marginata* reported by Mitoh and Yusa (2021) has recently gained a lot of attention, since these slugs have been shown to be able to decapitate their own heads from their body, which includes shedding of the whole heart, in order to rid themselves from a potentially parasite-infested body. This is an extreme case of *autotomy* (i.e. the behavioural strategy of deliberately shedding body parts), enabled by the special regenerative modes of being of gastropods. This is one way of responding to pathological complexity and thus entirely healthy behaviour. But it is also precisely in this context – in which behaviour is limited, and bodies are vulnerable yet allow for healing – that it makes sense to invest both in short-term states of pain and in longer term mood states such as fear or pessimism. Their life cycles last long enough to make such capacities a useful investment. However, we should not make too much here of the associations with certain rich human emotions and mood states. What we are interested in are these states as natural phenomena, which makes the human case a special case rather than a typical exemplar.

Due to the small nervous system that has made *Aplysia* a model organism to begin with, these results provide compelling functional evidence that a minimal degree of sentience may be present in these slow and vulnerable creatures. This view isn't anti-neural as much as it is gradualist. Because *Aplysia* belongs to the largest sea slugs, especially sea hares (*Anaspidea*) among them, such as the California sea hare (*Aplysia californica*), that are comparatively much more active – their movement resembling a "gallop rather than a slow crawl", as Godfrey-Smith notes – it can be hard not to grant them experience (2020b, p. 216). But despite their behavioural differences from smaller sea slugs which have very similar nervous systems, the opposition to attribution of sentience in the latter group may merely be a matter of perspective, with Godfrey-Smith arguing that once the smaller relatives are scaled up to the largest among the Aplysia, it becomes difficult to draw a hard boundary of experience, doubly so if their movement is sped up. A gradualist picture is tempting here and fits better with the actual data than the demand for a hard line. Even tiny slugs and their ancestors may possess an evaluative common currency of valence to deal with motivational trade-offs, despite a lack of capacities in the other dimensions. But to assess this idea, let us also examine another dimension in order to determine whether or not they really do have only evaluative experiences.

5.2.2 Sensory Experience

In the previous section I mentioned that gastropods, when compared to insects, seem to have fewer degrees of freedom in their behavioural repertoires. Furthermore, they have much simpler sensory capacities, though there are some exceptions. Godfrey-Smith (2020c), for instance, notes that sea elephants or heteropods (*Pterotracheoidea*) have something of a borderline case of type IV eyes, which might provide compelling evidence for sensory experience on the visual side. What also distinguishes the lifestyles of these species is that they are much more mobile – they have fins for free swimming and engage in predation, in contrast to most gastropods that live on the ground. The pathological complexity they are faced with is therefore different from the usual sluggish gastropod way of life.

For these swimming gastropods, with lifestyles more closely resembling the pathological complexity of fish and cuttlefish, we can make predictions regarding the likely richness in their sensory capacities. If sensory capacities are found in various degrees of complexity within a branch of life that is already a likely contender for minimal sentience, the pathological complexity thesis appears to gain striking support for the close relationship between lifestyle and experience. Most gastropods, however, appear to only have a "sliver of the features that make for experience in us" (Godfrey-Smith, 2020b, p. 262), and this sliver appears to be mostly on the evaluative side, providing compelling evidence for the independent existence of evaluative experience without strongly representationalist sensory capacities (or, for that matter, the other three dimensions). It is plausible that a transformation of this largely evaluative mode of subjective experience towards a richer representationalist form would simply not pay off for gastropods with their unique life histories, but that should not be taken as evidence that they do not have consciousness at all. As Ginsburg and Jablonka (2019) note, "in gastropod mollusks that have more complex brain ganglia than the sea slugs, the presence of elementary mental representation is an open question" (p. 394), but this will have to be investigated further.

Given the evaluative richness of gastropods, it is not at all implausible that they could also have a low degree of sensory experience. Godfrey-Smith (2020b) himself admits that gastropods appear to have a rich sense of smell (p. 215). In thinking about this dimension of consciousness, we appear to often be all too tempted to give priority to vision, which is precisely why I have replaced the term "perceptual richness" with "sensory richness" in my discussion of Birch et al. (2020). Nevertheless, since agreeing with Godfrey-Smith here is not detrimental to the pathological complexity thesis, let us now turn to the case of insects, which provides a stronger challenge. If Godfrey-Smith is right that insects possess sensory without evaluative consciousness, this could undermine my thesis altogether.

5.3 Arthropods: A Robotic Way of Life

Whereas Godfrey-Smith's (2020c) arguments for the presence in gastropods of evaluative experience without the sensory side provide strong support for the pathological complexity thesis, his arguments for the existence in insects of sensory experience without the evaluative side provide an interesting challenge that we will have to overcome. Godfrey-Smith (2020c) suggests that complexity in sensory "capacities might be understood as involving complexity in discrimination or in downstream processing" (p. 1153), but emphasizes the latter as being more important for considerations of subjective experience. While Birch et al. (2020) emphasized discrimination in sensory richness, Godfrey-Smith's emphasis on downstream processing is certainly reasonable due to a recognition of how many discrimination activities the brain is engaged in without subjective experience, even in humans. For the purposes of the discussion here, I remain neutral on the question of which side matters more, since the pathological complexity thesis sees the sensory experience as something operating within an evaluative sphere. Let us therefore look at the possibility that sensory experience could exist without such an evaluative space in which different sensory stimuli are being evaluated against each other.

Insects are part of the arthropod branch of life and constitute the great majority of arthropod species (in addition to that of all animals). They are estimated to have originated only roughly 479 million years ago during the early Ordovician, which suggests – as Misof et al. (2014) point out in a landmark study in Science – that they have evolved in response to the plants which started to colonize the planet around the same time (see also Labandeira, 2006). However, the arthropod group, which also includes crustaceans (e.g. crabs, lobsters, and krill), arachnids (e.g. spiders and ticks), and myriapods (e.g. centipedes), is a much earlier Cambrian invention – indeed, they constitute the paradigm phylum of the Cambrian explosion, leading the way for a special animal way of life. Their name, being a conjunction of the Ancient Greek words for "joint" and "foot", is a fitting description for a mode of being consisting of hard shells, multiple segments, and typically many appendages (Budd & Telford, 2009) that nevertheless shares a common active animal lifestyle with the "soft" and "sluggish" gastropods. But despite sharing a high degree of pathological complexity, it plays out differently in both groups and this might make it tempting to think that arthropods could evolve sensory consciousness without the presence of evaluation. To examine this further, this time we will begin with the sensory side of things.

5.3.1 Sensory Experience

Unlike the soft-bodied gastropods, arthropods seemingly overflowed in the Cambrian, with trilobites making up much of the fossil record. Partially, this is due to their possession of an exoskeleton, which simply fossilizes better, but their presence emphasizes much of the change that took place during the Cambrian. An exoskeleton makes *sense* as a protective shell against others, with appendages such as feelers and claws clearly existing in response to other subjects, whether prey, partner, or predator. Godfrey-Smith (2020b) describes arthropods as having "invented a new way of being an animal, with a skeleton that scaffolds and organizes complex actions. They also invented claws, and to go with them, image-forming eyes" (p. 80). The life histories of arthropods involve frequent interactions with others, so it is unsurprising that their sensory capacities and behavioural repertoire surpass that of gastropods – a richness that is also emphasized by Feinberg and Mallatt (2016). Indeed, if we think about the pathological complexity of insect life, it becomes surprising that few have granted them a minimal sense of subjective experience despite the prevalent vision-centric model of consciousness, since many insects have been shown to have sophisticated sensory capacities, especially high-resolution vision.

Notably, Godfrey-Smith (2020c) focuses on the much-studied bees and fruit flies (*Drosophila*), since it is here that we can examine flight as a complex behaviour that "involves dealing with complex spatial layouts and making self/other distinctions with respect to the causes of sensory events" (p. 1153). Indeed, in the ecological framework that I try to build here, flight constitutes the paradigm case for an explosion in pathological complexity, due to the rapid feedback between sensing and action. Godfrey-Smith (2020b) sees the sensory processing of flying insects as a plausible candidate for subjective experience, since it is such "feedback that contributes to a point of view" (p. 208). That flying creates a new challenge of complexity is not a new idea. In his very first publication, the British evolutionary biologist and former aeronautical engineer John Maynard Smith (1952) already emphasized the importance of a sophisticated nervous system in the evolution of flight, for both birds and insects. He argued that the evolutionary origins of flight must have required flight stability via a long tail, since early animals lacked the sensory richness and nervous system complexity to control such a flying body – similar to how pilots require a stable plane in order to be able to fly it. While such a tail lowers manoeuvrability, it greatly increases flight stability. Yet, Maynard Smith (1952) argued that "in the birds and at least some insects, and probably in the later pterosaurs, the evolution of the sensory and nervous systems rendered the stability found in earlier forms no longer necessary" (p. 129). The evolutionary advantages of unstable flight, he argued, would be the ability to turn more rapidly in the air and to be able to fly at slower speeds without falling (p. 128). Taking a design stance towards the problem of flight makes it obvious how rich the complexity of this problem really is. Free fall can mean death. But Maynard Smith made these comments in relation to birds, and insects are a different case.

Because insects are so small, air resistance will stop them from gaining enough fall speed to cause serious injury. Nevertheless, it is precisely because of their size that it is more important to focus on the organization of the insect nervous system and its decision-making speed, rather than the mere number of neurons. Broom (2014), for instance, notes that the critical flicker fusion tests have shown blowflies to have the fastest temporal processing speed, "perhaps not a surprise to all who have tried to catch one", and that they may "perceive humans as slow and inconsequential" (p. 118). Such findings could not only suggest that insects possess the sensory side of subjective experience, but also that it might be richer than in larger animals, including us – a conclusion we might even want to extend to their experience of time. Indeed, there is plenty of evidence that the more manoeuvrable and fast fly (or for that matter, bird) species are, the richer their temporal resolutions (Laughlin & Weckström, 1993; Potier et al., 2020), thus supporting the link between pathological and phenomenological complexity.

Yet, flying is in a sense a very new lifestyle and from an evolutionary perspective, we should think of flying insects as a distinct category, just as we treated the giant *Aplysia* as non-representative for the entire gastropod class. Their pathological complexity will be much higher due to their different lifestyles and in many ways resemble that of flying birds, which should lead us to engage in comparative thinking about the convergent evolution of phenomenological complexity. Here, I would also like to point out that we should give evolutionary thinking some priority over taxonomic thinking, which has historically been used mostly to claim that consciousness wouldn't exist in animals belonging to certain parts of the tree of life. Nevertheless, flying insects are themselves an evolutionary product and it is unlikely that their non-flying ancestors had poor sensory richness, since it is probable that at least a certain threshold of such complexity would be required. Though, following Maynard Smith, it could also be conceivable that the ancestors of flying insects had a long tail that provided a useful evolutionary scaffold for the evolution of flight, one that could later be discarded with the evolution of greater sensory capacities.[4] So while it makes evolutionary sense to attribute rich sensory consciousness to these insects, we should be careful not to necessarily attribute this capacity to all insects.

My aim here is importantly not to prove that insects have a sensory experience, only that it makes evolutionary sense within the context of the pathological complexity framework as an enrichment of evaluative capacities that became more representational. In order to address Godfrey-Smith's challenge that the sensory side could exist without the evaluative side, we will now simply assume that these animals are conscious, in order to evaluate whether such a separation makes sense for a conscious animal.

5.3.2 Evaluative Experience

One of the main reasons we should not think of sensory experience existing without evaluative experience is that it is ultimately through evaluation that richness on the sensory side "pays off". As Broom (2014) rightly acknowledges, the flicker fusion tests are closely related to the ability of animals to evaluate and make decisions, which is why he describes blowflies as having the "fastest evaluation" (p. 40). Yet, Godfrey-Smith does think that there is a compelling case to be made for a separation between evaluative and sensory sides of experience in insects.

In particular, Godfrey-Smith draws on an old but influential mini-review by Eisemann et al. (1984) in order to establish that "all known insects appear completely unconcerned about even severe body damage, despite having nociceptors. Wound-tending has never been seen in an insect, and after injury these animals just continue, as best they can, with the behavior appropriate to the circumstances" (Godfrey-Smith, 2020c, pp. 1153–1154). But this is partially a misrepresentation of even this early work on the possibility of insect pain. Indeed, Eisemann et al. (1984) cite early experimental work by Hentschel and Penzlin (1982) that showed grooming activity in response to damage and presented it as something to be *explained*. Eisemann et al. (1984) explicitly recognize an "increase in both general grooming activity and specific grooming of a wound site observed after experimental puncturing of the abdominal wall of the cockroach *Periplaneta americana* (L.)" (p. 166). While it is true that Eisemann et al. (1984) argue that insects do not feel pain, they do so in a very measured way, only stating that "the evidence from consideration of the adaptive role of pain, the neural organisation of insects and observations of their behavior does not appear to support the occurrence in insects of a pain state, such as occurs in humans" (p. 167). That they see the evidence as far from conclusive is also emphasized by their call to endorse Wigglesworth's earlier "recommendation that insects have their nervous systems inactivated prior to traumatizing manipulation. This procedure not only facilitates handling, but also guards against the remaining possibility of pain infliction and, equally important, helps to preserve in the experimenter an appropriately respectful attitude towards living organisms whose physiology, though different, and perhaps simpler than our own, is as yet far from completely understood" (p. 167).[5]

This then makes it somewhat puzzling as to why Godfrey-Smith (2020b) similarly repeats his assertion in *Metazoa* that "insects have still never been observed tending and grooming injuries; that claim from the old no-pain paper still holds up" (pp. 211–212).[6] Just because insects have not been shown to engage in sophisticated "protective behavior towards injured body parts, such as by limping after leg injury or declining to feed or mate because of general abdominal injuries" (Eisemann et al., 1984, p. 166) does not mean that no grooming-like behaviour has been observed *even if* it could be explained in a way unrelated to pain. The way Eisemann et al. (1984) deal with Hentschel's observations is to point out the "contra-adaptiveness of this response in relation to wound healing" (p. 166). But we have to distinguish the adaptive value of such behaviour from its support for the presence of subjective experience. It may very well be the case that not all grooming behaviour is adaptive, no less so than the scratching of human wounds is. Pain could be invoked as a cause as long as a general experienced negative valence exists regarding damage or potential damage. Indeed, this might even be seen as supporting the presence of negative valence as opposed to a mere "mechanical" response.

My argument here, however, should not be read as me endorsing the presence of pain in insects – this is yet to be established. I only argue that the case is not as straightforward in insects as Godfrey-Smith makes it seem. Nevertheless, it is certainly true that insects – more so than perhaps any other complex agent-like animal group – have been observed to be apparently oblivious to all kinds of bodily damage and injuries. They engage in sex and feeding while being devoured, soldier on despite damages, and even eat their own insides when they are leaving behind their body due to damage (Adamo, 2016; Eisemann et al., 1984; Walters, 2018). In order to understand whether or not such behaviour is functional, we will have to understand the pathological complexity faced by insects. Godfrey-Smith (2020c) notes the "ecology of insects is also relevant" (p. 1154), but for a true cognitive ethology it should be our primary source of information when extending the Darwinian revolution to consciousness. The references by Godfrey-Smith (2020c) to the life histories of insects versus crustaceans are particularly interesting here.

Whereas most crustaceans live in the water, leading similar lifestyles to their Cambrian ancestors, insects have predominantly branched towards a life on land.[7] Yet, whereas Godfrey-Smith wants to deny evaluative experience in insects, he grants it to crustaceans, where wound-tending has been shown (Birch et al., 2021). The work of Elwood and his collaborators (Appel & Elwood, 2009; Elwood, 2012) has been particularly influential in this context, which – similarly to the early work by Cabanac – studied the evaluative trade-offs decapod crustaceans (shrimps, crabs, and the like) are engaged in. For instance, hermit crabs have shown that they are making state-based decisions on whether or not to leave their shell when receiving electric shocks, dependent upon both the predicted presence of predators and the shell value. In a review for the Department for Environment Food and Rural Affairs (Defra) of the UK that covered a vast range of further evidence – so vast in fact that it could fill an entire chapter – Birch et al. (2021) recently argued that decapod crustaceans are sentient. This has led to them being added (alongside cephalopod molluscs) to the list of animals included in UK animal welfare legislation. So it appears plausible that the evolutionary ancestors of these animals in the Cambrian already possessed sentience.

If we grant this, however, then this admittedly transforms Godfrey-Smith's challenge to the pathological complexity thesis. Instead of sensory experience arising independently in its own right, the challenge now appears to be explaining a *loss* of the evaluative side in insects once the sensory side has come to play a more important role. After all, these results are convincing enough to have motivated Tye (2016) to call his book *Tense Bees and Shell-Shocked Crabs: Are Animals Conscious?* Allen and Bekoff once criticized as anti-neural Griffin's suggestion that bees might have more of a use for subjective experience *because* their nervous systems are so small (1997, p. 153), but I take it that Griffin's motivation was to highlight the important insights from the ethologists to emphasize the lifestyles of animals prior to conducting laboratory experiments.

Within the peculiar pressures of life on land, most insects have evolved very short and routinized life histories that differ from the comparatively longer and "less regimented lives of their marine relatives studied by Elwood" (Godfrey-Smith, 2020c, p. 1154). While there are exceptions to this rule (Maruzzo & Bortolin, 2013; Suzuki et al., 2019), insect limbs generally do not regrow and there is little evidence that there is adaptive value for them in protecting injuries since their capacities for healing are quite poor. Godfrey-Smith (2020c) describes this lifestyle as being about *soldiering on* even in the face of pathologies (p. 1154). Now, this makes sense in a semelparous lifestyle and especially in eusocial insects, where individuals are replaceable. One might expect bees or ants to have sophisticated sensory capacities for finding food sources but to be less rich on the sensory side *in order to* focus on their tasks. But does this really show that the evaluative side has been lost?

Godfrey-Smith (2020c) admits that bees have been shown to avoid noxious stimuli such as heat, but notes that this could be a mere reflex, not necessarily involving subjective experience. A compelling line of evidence in this context are various forms of learning ability, since they are commonly taken to constitute convincing evidence for the possibility of evaluative experience. Unfortunately, the abilities of insects to learn are routinely underestimated by both scientists and philosophers. Figdor (2020), for instance, criticizes Dennett's (1973) discussion of the Sphex wasp – "characterized as an inflexible machine incapable of any learning" as a clear case of "confirmation bias" (p. 2). But there is no longer a debate on whether insects are capable of reinforcement learning (Allen et al., 2005; Elwood, 2012), and insect consciousness is starting to be taken as a serious possibility. Which forms of learning constitute the best kinds of evidence for consciousness, however, remains contested. Ginsburg and Jablonka (2019) provide a good overview of this debate and argue that there is unlimited associative learning that provides something like a transition marker that animals are conscious, though they do not mean to say that the absence of unlimited associative learning necessarily shows that consciousness isn't present. I am very sceptical that we can actually find anything like a definite marker, since consciousness can come in a diversity of forms, but that is not an objection to the idea that sophisticated forms of associative learning constitute good indicators at least for a certain richness or even transition of consciousness, as opposed to its presence. But as with gastropods, we should also highlight evidence for nociceptive sensitization as indicative of evaluative richness in insects – which is notably also highlighted by Tye (2016). One peculiar result that Godfrey-Smith (2020c) emphasizes is the presence of sensitization in *Drosophila* larvae, as opposed to its later life stages (p. 1156). Too much focus, he notes, might have been given by Eisemann et al. (1984) and Groening et al. (2017) to the absence of pain in adults:

> Another factor in insects not highlighted so far, one related to life on land, is the differences between larval and adult states. Many insects lead two lives, in effect, one on each

> side of a metamorphic divide, with extensive breakdown and reconstruction at that stage. In the kinds of insects considered here, it is the adult who has acute sensing that controls complex motion; the larva does not.
>
> *Peter Godfrey-Smith (2020c, p. 1156)*

Drawing on Sprecher et al. (2011), Godfrey-Smith (2020c) emphasizes that larvae have very simple eyes – in *Drosophila* only the small number of 12 photoreceptor neurons dedicated for vision, much simpler than in the adult stages. Yet, in contrast to the apparent obliviousness to damage in adult insects, Godfrey-Smith (2020c) also highlights the work of Walters (2018) that showed larval stages of *Manduca* and *Drosophila* to have nociceptors and nociceptive sensitization. What we find in insects is a striking disconnect between the pathological complexity faced by the larval and adult stages.

Partially, such observations are due to the extreme diversity of insect life cycles, which explains the presence of a huge variety of alternative life-history strategies, including such odd examples as male mantids engaging in sex with females despite being eaten afterwards. While this behaviour may well be adaptive (Schwartz et al., 2016; Zuk, 2016), it is hard to think about such extreme behaviours as involving pain. And yet, larvae – despite their nervous system simplicity – often appear to have richer evaluative capacities than adults, indicative of the different emphasis on survival during this stage, as opposed to reproduction in the adult one. The adult insect body is described by Godfrey-Smith as a mere tool for this end. So it would make sense to have life-stage-dependent varieties of experience. This is something that can straightforwardly be captured within the pathological complexity framework, providing us with different measures at different stages of a life history.

Barron and Klein (2016) argue for insect consciousness, but largely emphasize the sensory side of things. Yet, findings on the evaluative side are compelling. Godfrey-Smith (2020c) points to the self-administration of analgesics which has been used as compelling evidence of pain in birds and fish, yet has not been found in bees (Groening et al., 2017). But these and other studies have been challenged (Gibbons & Sarlak, 2020), leading Browning and Birch (2022) to urge us to be cautious since there is little evidence to expect that morphine must be "a good analgesic in insects" (p. 5). But that is not the only source of evidence we can look for. Bateson et al. (2011) show convincingly that bumblebees, if they have been shaken, can have negative long-term mood states called pessimistic bias. Similarly, Godfrey-Smith (2020c) admits that bees and other insects show aversive responses to heat, which may be a better stimulus with which to look for the presence of subjective experience. As I have argued elsewhere, even if insects do not experience human-like pain towards mechanical injuries, they may very well experience other aversive experiences such as hunger or thirst (Browning & Veit, 2020b). The absence of pain is too often confused with the absence of evaluation. But the lifestyles of insects simply don't make it necessary to put much of a value on protecting one's bodily shell from mechanical damage. What is important is to complete one's life-history strategy: i.e. to reproduce (or in eusocial insects, to help your sisters with the said goal). If wound-tending doesn't aid that, there is little sense in putting much valence on it.

Instead of focusing on pain-like behaviour as an admittedly tempting but flawed paradigm case of evaluative experience, we should look at the evidence for a valence system more generally, one that evaluates conflicting stimuli in a flexible manner. Evidence that is very compelling here, and also highlighted by Godfrey-Smith (2020c), is a follow-up study to Bateson et al. (2011) which focused on positive mood states in the form of optimism bias in response to unexpected sugar rewards in bees (see Perry et al., 2016). The idea of pleasure as a common currency is sometimes criticized as failing to account for the different neural mechanisms of negative and positive evaluation, but such common functional roles of the evaluation suggest that they are deeply evolutionary intertwined. Indeed, they *must* largely operate in tandem to allow for efficient decision-making and learning in the face of novel and ambiguous stimuli.

Given how much we know about the level of sensory processing taking place unconsciously in human brains, it is plausible to think that it is only those sensory inputs that enter the affect system at the final common path of the brain that are consciously experienced (see Ginsburg & Jablonka, 2019 for a similar view). Perhaps a better description would be to see the evolution of "sensory consciousness" as an enrichment of the evaluative side with respect to its discriminatory capacities. Such a view provides us with an answer to those who insist that functionalist accounts of consciousness cannot explain the "feel" of experience, since it is precisely this subjective experiencing that does the functional work of dealing with motivational trade-offs; it enables organisms to efficiently deal with their species-specific pathological complexity. So while we could readily admit that insects do not feel pain due to their "robotic" way of life, their complex behaviours and learning abilities are highly suggestive of something like a common currency of valence for efficient action selection, even if their evaluative capacities on this side of things may have become poorer compared to their sea-living ancestors. Insects, after all, are in many ways the scaled down versions of their Cambrian ancestors, with a constant pressure for efficiency in organization, especially in those insects that can fly.

Yet, despite (or perhaps precisely because of) this size constraint, the nervous systems of bees have been shown to be incredibly complex. It is because of this that they might provide us with an insight into the minimal nervous system requirements for sentience. Lack of evidence should not here be confused with evidence of absence, precisely because dedicated research on the evaluative capacities (as opposed to their sensory capacities) has been rare. But this is now beginning to change. Indeed, on the 26th of July 2022, a particularly compelling article on motivational trade-offs in bumblebees has been published (see Gibbons et al., 2022). In it, the authors showed that when faced with noxiously-heated feeders and different sugar concentrations, bees were able to trade off competing motivations which they argued could be suggestive of insect pain. Indeed, their work suggests that bumblebees do have a common currency for evaluation. Lastly, recent reviews of insect sentience offer ample additional support for their subjective experience of evaluation (see Lambert et al., 2021; Mikhalevich & Powell, 2020).

In conclusion, even what is supposedly the best case for the independent existence of sensory experience without evaluative experience – insects – turns out to demonstrate rich evaluative capacities after all, thus supporting the motivation of the pathological complexity thesis to seek the origins and core of subjective experience in hedonic valence. As Godfrey-Smith (2020b) himself recognizes, it may very well be true that adding a new dimension of consciousness on top of an older more fundamental one may radically transform the phenomenon itself, shifting older features into the background, but that doesn't mean they have disappeared. It is certainly obvious from the human case that the evolution of rich vision can dominate consciousness and we should happily embrace the possibility that subjective experience is as diverse as the diversity of alternative forms of life we find in nature. After all, it is this diversity that ultimately motivates the link between pathological complexity and phenomenological complexity. Let us therefore now think about the other three dimensions and how they may transform consciousness.

5.4 Octopuses: A Disembodied Way of Being

Just as insects challenge our usual thinking about the centrality of pain in evaluative experience, and gastropods challenge our thinking about the need for a sensory experience in the conscious being, octopuses constitute a challenge to our usual thinking about self-consciousness. What is it like to have a body that can be moved in (figuratively) almost any shape one could want? Their peculiarity and "alienness" has gained them plenty of attention in discussions of animal minds, and animal welfare legislation in many countries now includes them as "honorary" vertebrates on the list of species deserving protection. Even the early paper by Eisemann et al. (1984) that denied insect pain already suggested that "more caution would be needed in other cases, notably that of the cephalopod molluscs, which have a considerably more complex nervous system" (p. 167).[8] Among the molluscs, octopuses stand out for their unique life-history complexity arising from their high degrees of freedom. Furthermore, their soft bodies and foraging lifestyles make them susceptible to predation and parasites, driving up their pathological complexity. Giving up any kind of shell to protect themselves, in addition to possession of an incredible amount of cognitive sophistication and behavioural flexibility, has perhaps made them the most interesting case for an independent evolution of a rich kind of self-consciousness. Like insects, octopuses have short lives, and yet, their bodies appear to be rich and vulnerable investments, rather than the armoured-vehicle-type bodies of insects. Indeed, this combination almost makes them a paradigmatic case for the pathological complexity thesis; vulnerability paired with a high investment requires some sort of conscious control and agency to ensure that this risky use of resources pays off.

Other molluscs might be different in this respect. Walters (2018) notes, for instance, that whereas octopuses have been shown to engage in wound-tending (Alupay et al., 2014), evidential support for the same is lacking in squid, though he also notes that this may simply be an artefact of a lack of studies since "few scientists working in pain-related fields have investigated squid" (p. 7). Nevertheless, there is plenty of convincing evidence that octopuses experience pain-like states, with specialized sensory neurons which appear to focus on noxious stimuli (Alupay et al., 2014; Crook, 2021). Here, I am less interested in arguing for the presence of consciousness in these strange but beautiful animals. This case, I think, has been successfully made on the evaluative and sensory sides in recent years. Despite being colour-blind (Bellingham et al., 1998; Marshall & Messenger, 1996; Messenger, 1977), Birch et al. (2020) describe them as having "rich visual and chemo-tactile perception" (p. 796). Furthermore, following Birch et al. (2020), Mather (2021b) has also written a dedicated article on sensory richness in octopuses, so there is less reason for me to do so here. Likewise, the aforementioned Defra report by Birch et al. (2021) supports a view of very rich evaluative capacities in cephalopod molluscs, including nociception, grooming, and associative learning, among other indicators for evaluative experience. While we could go into much more detail on these two dimensions, this section will instead use octopuses to challenge our preconceived ideas about the dimension of self-consciousness.

5.4.1 Self-Consciousness

Admittedly, I feel some unease in discussing self-consciousness and the unity of experience separately from each other. As I have argued in Chapter 2, these dimensions are incredibly closely related, but since my arguments here do not rely on them being necessarily separate, it won't matter all too much. Octopuses constitute an interesting natural experiment not only for a very different kind of selfhood but also for the possible synchronic disunity of mind. They have often been discussed in this context since their cerebral ganglia and brachial plexus (the nerve ring connecting the neurons in the arms) are arranged in a way that allows some degree of functional autonomy (Carls-Diamante, 2017; Ginsburg & Jablonka, 2019; Godfrey-Smith, 2016d, 2019b, 2021a; Mather, 2019). They appear to have a degree of agency in each arm rather than transferring all information to some sort of central agential control.

While they have both lateralization and centralized integration, their nervous system is best described as *another* strikingly different way of being, perhaps constituting something like a switching of control between arms and the whole organism, rather than two distinct subjects (Godfrey-Smith, 2016d). Other admittedly bolder options would be to think about an octopus as having nine selves – with each individual arm having their own consciousness, their own subjectivity that is only loosely integrated with a higher-level subject: nine selves "housed" within a single body. Can there be something like a theatre of subjects with more or less unified experiences? Do they all have their own perspective or point of view or is this a case where this metaphor breaks down? After all, as Birch et al. (2020) point out: "Octopuses prefer different arms for different activities and their arms appear to function partly independently of the brain" (p. 796). Our folk understanding of consciousness seems to resist such notions, but this may not reveal anything important about the core features of consciousness as a phenomenon in nature.

Since the nervous systems of insects and gastropods may well be sufficient for sentience, we should not be surprised if there could be numerous selves in octopuses, with each of their arms being engaged in their distinct trade-off optimization problems. Within the pathological complexity thesis, this would be a sensible prediction to make and we should conduct more empirical work on "arm-cognition" and information transfer between them to test this hypothesis. We currently lack the evidence to make a confident assertion in either direction, but this is precisely why a comparative approach is needed in order to learn more about the role of *selfhood* and *unity* in consciousness. Importantly, much of the information processing in octopuses is divided between either right or left parts of their brain (Schnell et al., 2016). But because Mather (2021a) has recently offered a paper discussing synchronic integration in octopuses, it will be doubly useful to focus instead on the experience of a self. Let us thus move through the "levels" of selfhood and think about their possible evolution by using the lens of the octopus.

5.4.2 A Minimal Sense of a Bodily Self

The evolution of a minimal kind of selfhood, as I argued in Chapter 4, is plausibly found in an enrichment of the sensory side. If consciousness begins with a minimal feeling of evaluation, and later becomes representationally enriched to discriminate more stimuli, it is not at all hard to imagine how further enrichments of just this sort would lead to the feeling of a self in virtue of an eventual distinction between interoception and exteroception. There does not appear to be a further hard problem here.

What is important to recognize in discussions about the gradual evolution of consciousness and currently existing animals is that many of the gradual steps between the transitions we see today plausibly went extinct. Future evidence may well show that gastropods and insects have a minimal bodily self. Such findings should not be taken as evidence against the pathological complexity thesis. Once an evaluative "core" of subjective experience appeared in the Cambrian, it is likely that arms races would quickly have led to representational enrichments of this capacity, allowing for better action control. An awareness of oneself in space would quickly have become useful, similar to the tracking of one's total bodily state. The reason the *state-based* life histories of these animals are so difficult to model is precisely because different bodily states demand very different actions, and any animal that is able to flexibly alter its behaviour would be able to reap unexplored teleonomic benefits.

Further, I also noted that there is a problem with distinguishing conscious from unconscious self-perception. However, the pathological complexity thesis can here again be used as an answer, by emphasizing only those sensations that enter into the evaluative sphere at a final common path. Most of what goes on in human self-recognition, e.g. balance, walking, and the like, is typically not consciously experienced. It is only under pathological conditions (e.g. damage to the ears), or slippery/moving surfaces (e.g. riding a horse), that our position in space enters our subjective experience, forcing us to consciously balance. But where animals have a very flexible body, we can predict that selfhood will play a more important role in centralized decision-making. Octopuses are able to distinguish their arms from other objects, precisely because their arms could otherwise get in each other's way, which perhaps implies a self-recognition mechanism in each arm to minimize interference (Nesher et al., 2014). Most of this is plausibly unconsciousness, but when conflicts in information from different arms arise, there will likely be a need to consciously "disentangle", and this strongly supports the presence of "embodied" experience. A minimal degree of this dimension can be expected in all animals with a sufficiently high degree of sensory consciousness.

5.4.3 Self-Awareness

Birch et al. (2020) cite the work by Boal (2006) on social recognition in cephalopods to note that they probably do not possess self-recognition, due to their more or less "anti-social" lives. Yet, Birch et al. (2020) also cite work on their extraordinary camouflage skills that enable them to hide and pretend to be inanimate objects (Hanlon, 2007; Hanlon et al., 2011; Huffard, 2006) or imitate noxious animals (i.e. "fish mimicry"; Hanlon, 2007; Hanlon et al., 2010) or clumps of algae (Josef et al., 2012). So the current evidence base may be indicative that the flexible implementation of such camouflage tactics suggests at least some understanding of their own bodies and how others experience them.

Here, it is important to pay very close attention to the life histories of the animals to design tests that mimic their typical lifestyles. Given that octopus bodies are so vulnerable and changeable, it is natural to think that their "body image" is not anything like the stable perception we have of ourselves in space. If a breeze of wind was able to lift up our arms and make them flutter in the wind, we may be better able to conceive what it is like to have an octopus body, but I do not want to speculate too much here (though see Levy & Hochner, 2017). The point is merely to recognize that it is our comparative embodied-ness that is typical for most conscious beings, but that may simply not extend to octopuses in quite the same way (see also Godfrey-Smith, 2016d). Where many animals would see a hole and be able to conceive whether they fit through or not, this question is a bit of a trivial matter for an octopus who can fit through a hole much smaller than the size of their body spread out on the ground. What is the sense of having the capacity for self-awareness in the absence of an adaptive rationale? The conflicting evidence is suggestive of perhaps something in between our usual thinking of self-awareness and minimal embodied phenomenology, but there is no need to think that this would be an experience very similar to our own or that of other mammals. It might very well be a very alien kind of subjective experience.

5.4.4 Theory of Mind

In thinking about the evolutionary role of the experience of a self, an important function might be the inference from oneself to others, rather than a better organization of one's own sensory inputs. Birch et al. (2020) highlight a recent false-belief task experiment by Kano et al. (2019) that involved orangutans (*Pongo abelii*), chimpanzees (*Pan troglodytes*), and bonobos (*Pan paniscus*) in a test involving both a translucent and an opaque barrier. Interestingly, only those apes that had themselves experienced the barrier as opaque inferred that others would not be able to see objects on the other side of the barrier, thus highlighting the important role of *self-experience* that both human infants and other animals rely on to predict the behaviour of and attribute mental states to others (see also Meltzoff, 2007). This research on self-experience is important because it suggests that "animals can make experience projections, inferences from what they experience in a particular situation to what others will experience" (Birch et al., 2020, p. 797). In addition, there are interesting findings of mindreading abilities in corvids who have been shown to engage in what Birch et al. called *experience projection*, i.e. the prediction of others' behaviour based on their own experience in a similar situation (Dally et al., 2005; Emery & Clayton, 2001; Ostojíc et al., 2017). Such experiments may be harder to design for octopuses and other sea animals, but that obviously doesn't mean that they shouldn't be attempted, since they may help us to better understand the theory of mind capacities in other animals. Wherever there are complex decisions to be made that involve predicting the knowledge of others, it is plausible that these will enter the subjective experience of animals.

Here, it is useful to highlight a recent target article titled "Knowledge before Belief" in *Behavioral and Brain Sciences*, where Phillips et al. (2020) propose a radical reversal of the current paradigm in "Other Minds" research. Breaking with a long tradition that sought to understand the minds of other humans (and animals) by focusing on the attribution of beliefs, the authors argue that decades of empirical research in the cognitive sciences have undermined, or at least begun to call into question, the assumption that the attribution of knowledge rests on a more basic or fundamental capacity to attribute beliefs. For historical, methodological, and philosophical reasons, however, other minds research has long been held back from even considering this option within the conceptual space.

There is a long-held popular view in philosophy to analyse knowledge as justified true belief, with belief thus constituting a more basic concept (Ichikawa & Steup, 2018). But such a conceptual analysis may not be very convincing when overwhelming evidence from the cognitive sciences appears to tell a different story. Whereas even great apes have consistently been shown in a number of studies to fail at a broad range of false-belief tasks (Krachun et al., 2009; O Connell & Dunbar, 2003), Phillips et al. point out that they nevertheless succeeded at representing the knowledge of others (Hare et al., 2000; Karg et al., 2015; Melis et al., 2006; Whiten, 2013). If Phillips et al. (2020) are correct, we would be getting a more fine-grained distinction between the representation of others' beliefs (a trait that may be unique to humans) and the representation of others' knowledge (a trait that may be evolutionary ancient). Focusing on human representations of other minds might have biased us once again against a much more basic approach to other minds research.

In a reply to Phillips et al.'s work, I have argued that the ability to track others' knowledge is an evolutionary ancient trait appearing roughly at the beginning of the Cambrian (Veit, 2021b). What is notable in the early Cambrian is an increase in body size and the emergence of various sensory modalities to track one's environment. But more sophisticated ways of sensing one's surroundings naturally led to ways of sensing others – to react. This emergence of a richer kind of agency gave rise to arms races between predators and prey (Bengtson, 2002) and the evolution of centralized nervous systems (Wray, 2015) to better coordinate action and perception.

An important observation made by Godfrey-Smith (2016a) is that there is a transition somewhere in the Cambrian after which "the mind evolved in response to *other* minds" (p. 63). This transition should be understood as the evolution of representing the knowledge within other minds, which can lead to another (though admittedly smaller) explosion in pathological complexity. An important question for both predator and prey becomes: "*Have I been seen?*" The existence of eyes appears to function as a shorthand for many animals to make just this inference – when eyes meet, one infers knowledge of one's location to the subject at the other end of this exchange. Burrowing, camouflage, ink release, and flight are useful attempts to break this link. Many predators avoid the eye contact of their prey at all cost. Knowledge and ignorance of one's surroundings can make all the difference to survival. The evolution of eye-spots on butterflies is one such invention to make potential predators think that they are seen, thus avoiding conflict. Behaviourists may appeal to simpler explanations, but in this case, knowledge attribution may not be such a complex affair.

While octopuses do not engage much with each other, they are active hunters and a desirable source of nutrients to predators. This peculiar life history – that of a lack of any protective shell, instead involving lots of hiding, camouflage, and tool use (Finn et al., 2009; Hanlon et al., 2010) – is indicative of a special awareness of one's surroundings. The hide-and-seek activities of octopuses at least make sensible the possibility of attributions of an understanding of others' experience. While subjective reports of divers and researchers engaging with octopuses require empirical validation through further experiments, this does not mean that they should not be taken seriously. As Browning (2017) argues, anecdotes can be evidence too and there are plenty of reports of octopuses recognizing different humans, thus suggesting that there is much to learn about these other minds. Some capacities for the recognition of others are likely extremely widespread in the animal branch of life, including octopuses, even if these capacities are less sophisticated than in us. Nevertheless, octopuses will present a continuous challenge in our thinking about self-consciousness, with much further work remaining to be done. Their peculiar bodies and lifestyle make them as Schnell and Clayton (2021) argue "suitable ambassadors for rethinking cognition" (p. 27). Let us therefore now move to an exploration of the dimension of synchronic unity, as represented in fishes and non-avian reptiles.

5.5 Fishes and Non-Avian Reptiles: Disunified Ways of Being

In Chapter 3, I mentioned that Birch et al. (2020) suggest using birds as natural experiments of split-brain patients. Yet, they overstate the similarity to split-brain patients since birds have the lower parts of their brain connected, which also makes up a larger fraction of the whole than the cortex does in us (Vallortigara, 2000). One would therefore be too quick if one simply treated them as a case for two subjects within a single body – they are more of an intermediate case. Non-avian reptiles – or for that matter, fish – constitute the best neuroanatomical examples of a healthy split-brain because they have (compared to birds) greatly diminished ipsilateral projections in the tectofugal and thalamofugal pathways, which make it easier to infer that only one brain "half" is involved in a task (Deckel, 1995, 1997; Sovrano et al., 2001; Vallortigara, 2000).

Indeed, the evolution both of birds and mammals[9] is a striking case of an increase in lateral brain connectivity, which is precisely why we should be careful not to overemphasize birds. Their evolution from theropods (the clade of dinosaurs including all predators) is one of consciousness becoming more unified. In thinking about healthy and adaptive forms of disunity, we should therefore focus on fish and non-avian reptiles, which I put here together not because they have very similar life-history strategies, but rather because they remain comparatively understudied in comparison to octopuses and birds.

Notably, despite a long history of disregard of the cognitive capacities of fish and non-avian reptiles due to their status as vertebrates, there is now a broad consensus that they are conscious. There are, of course, still those who contest the presence of pain in fish (see Key, 2016) and discount the cognitive abilities of non-avian reptiles (see De Meester & Baeckens, 2021), but if we want to think about the evolution of synchronic unity within the context of the pathological complexity thesis, these groups constitute the best test cases since we can assume basic evaluative, sensory, and selfhood-related capacities. For excellent reviews of their abilities and capacities linked to consciousness, see Braithwaite (2010), Allen (2013), Woodruff (2017), and Sneddon et al. (2018) in the case of fish and Lambert et al. (2019) and Learmonth (2020) in the case of non-avian reptiles. But let us now turn to the capacity for synchronic unity.

5.5.1 Synchronic Experience

If many researchers think that human split-brain patients house two minds, shouldn't we also think of these healthy split-brain organisms as having disunified experience? Could it be an efficient way of dealing with the pathological complexity faced by snakes, lizards, and fishes?

As I have argued in Chapter 3, I do not think of the dimensions of unity as genuine ingredients of consciousness. They are rather properties consciousness can have (or, for that matter, lack), and I thus see little reason for thinking that they constitute anything like an additional explanatory gap. A certain degree of unity is an automatic feature simply in virtue of informational bottle-necking and the fact that multicellular animals typically act as integrated wholes. A discussion of the evolution of unity (or disunity) must focus on gradual increases or losses in said capacity, not a binary distinction between unity and disunity. Consciousness is highly integrated in us, but that is no reason to think that it must be so in other animals such as fishes and non-avian reptiles, who have very lateralized brains and division of labour between the hemispheres. Does this mean they have two final common paths? Not quite, since the performance of most behaviour is by necessity ultimately excluding other behavioural options. Yet, cooperation between brain halves might be a good way to think about lateralized brains. An analogy to tennis players by Godfrey-Smith (2021a) is useful here: just because two agents may work together to form a team, or even something like a collective agency, does not mean that their own perspectives blend into one subjective experience (p. 293). Importantly, not everything that seems to go on in one of the hemispheres is ultimately communicated to the other. There are actual gaps it seems, making it an interesting question to ask why this isn't pathological as it is in the human split-brain case.

Yet, the very question is loaded. To think that lateralized brains are an inferior way of being is a bold assumption, given that the highly lateralized brains of fish and non-avian reptiles are a product of natural selection (i.e. their brains have become more lateralized during their evolution) implying that there must be certain efficiency gains from having a lateralized brain. That human-centric approaches to consciousness may suggest that animals become less conscious through the evolution of lateralization should be counted against those theories, not as evidence for the lack of consciousness. It is a striking fact that vertebrate evolution seemed to have very little selective pressure for *corpus callosum*-like connections between the upper brain hemispheres, which leads others like Godfrey-Smith (2021a) to similarly conclude that such a lateralized way of being can be advantageous. From a neuroeconomics perspective, this should admittedly not be all too surprising. There is no way an efficient system can be built (whether biological or artificial) that can make complex decision-making calculations by bringing all information through a single bottleneck. The system would be far too slow, and in the animal case this would mean death. The philosophers' emphasis on agency and goal-directedness routinely leads to the temptation to think of a perfect mind as a Cartesian theater, and we must continuously resist this urge.

Interestingly, Vallortigara and Rogers (2020) follow up this kind of reasoning to argue that the exploitation of asymmetries allows for efficiency gains in vertebrates with lateralized brains. If this is the case, it would explain the preferences of animals for doing particular tasks with one side of their body – not as a mere preference of liking, but as a deep evolutionary efficiency rationale. Drawing on work by Babcock and Robison (1989), Vallortigara and Rogers (2020) argue that lateralization already took place in the Cambrian, when ancestral arthropod-like predators such as *Anomalocaris* "were preying on trilobites with a right-limbed bias" (p. 275). If we were to add into life-history theory such game-theoretic factors in species interactions, pathological complexity would dramatically increase – giving us an explanation for why such biases may evolve even in the absence of direct efficiency rationales for the organism. Their adaptive nature might only be understood once we pay close attention to the ecological lifestyles of these animals, which include knowledge of their predators and prey. Perhaps, a high degree of pathological complexity positively calls for a "splitting" of consciousness into separate streams.

Godfrey-Smith (2021a) is happy to endorse the idea of two separate streams of cognitive or sensory capacities, but when it comes to the affective side, with evaluative states such as fatigue, he maintains that they are likely shared across the hemispheres – thus giving us partial unity. Regarding the apparent difficulty of imagining such a partial split of experience, he attributes to our lack of imaginative capacities, whereas I think these cases are pretty easy to think about once we pay close attention to the life histories of these animals.

Recall our discussion in the preceding chapters of electroencephalograph studies of sleep, which show that animals can rest one brain hemisphere while the other stays active. Does it make sense to think of "tiredness" as being felt on both sides when the "restfulness" of each hemisphere can be distinguished? To me, it seems that in such cases there is good reason to think that tiredness could exist separately in both brain halves and lead to sleep in one half when it becomes too tired. Imagine a scenario with high predator load and preferential attention to predators given on one side. Does it not make sense that this side would tire more quickly and that the organism would keep track of that? Other states like hunger may be different since they are better conceived as whole-body states, but the argument for the existence of disunity isn't that every part of experience must be disunified. Animals with disunified brains may be able to flexibly merge their two streams of experience, if they exist, to allow for better coordination, or try to break them apart for better efficiency in particular tasks.

These are inherently speculative suggestions, but the more we learn about split brains in humans and animals, the more realistic these scenarios become. By paying close attention to the life histories of these animals, I hope that future work will enable us to better understand the unique adaptive benefits and drawbacks of the integration of experience. As I hope to have demonstrated, simply thinking about the lives of these animals enables us to gradually move away from the human-centric bias in thinking about consciousness. What is pathological in humans may offer important adaptive benefits to the life histories of many animals. Much research remains to be done on synchronic unity in these understudied animals, but little sense remains in asserting that brains must be highly integrated, as opposed to lateralized, in order for consciousness to exist. With this in mind, let us turn to the last dimension, i.e. diachronic unity.

5.6 Corvids: A Cunning Way of Being

Few animals have been given as much attention in the study of animal cognition as birds. This is unsurprising for two reasons. Firstly, birds are highly intelligent and can often learn experimental procedures quicker than other animals can – thus making them ideal partners in the lab for the study of non-human cognition. Secondly, and perhaps more importantly in the context of this book, the observation of birds was a major cause of both the Darwinian revolution (think of Darwin's finches) and the extension of the Darwinian revolution through the ethologists such as Lorenz (think of his geese), so it is perhaps unsurprising that their study is now also heavily implicated in trying to extend the Darwinian revolution to the mind. As a bird-lover, I give the last substantive discussion in this book to them, which I had the pleasure to meet during my time in Nicola Clayton's comparative cognition lab at Cambridge – or to be more precise the group of birds called the *Corvidae* including diverse groups such as crows, jays, magpies, jackdaws, and ravens.

I have already in the discussion of insects emphasized the explosion of pathological complexity with the evolution of flight, and this will be a much more important factor in birds due to their size. Free fall here means death and this may be able to explain the difference in how we treat animal cognition and sentience in birds in comparison to non-avian reptiles. Indeed, corvids appear in many ways akin to us making them very different cases of non-human consciousness than, say, octopuses. They are long-lived and their life histories are typically highly social, involving the recognition of the mental states of others (Keefner, 2016), similarly also to apes, which makes it hardly surprising that they have a high degree of cognitive complexity. As Schnell and Clayton (2021) acknowledge: "Several current views on the evolution of complex cognition suggest that intelligence coevolved in animals with slow life-histories in response to key socio-ecological challenges" (p. 29). But what makes them particularly interesting in discussions of diachronic unity are their incredible memory capacities, offering us an excellent target system to study the experience of time.

5.6.1 Diachronic Experience

Corvids have been one of the paradigm cases to study mental time travel (i.e. the memory of past experiences and the simulation of future scenarios/experiences), due to their food caching activities. In the longer life histories of birds, caching makes sense, since food can be stored for future periods such as winter where it might be scarce, thus providing a significant survival benefit against the pathological complexity of bird life, despite requiring an investment into (presumably costly) memory capacities. Clayton and Dickinson (1998) have called the capacity of scrub jays to remember the what, when, and where, of their caching events "episodic-like memory", which is meant to be distinguished from episodic memory since episodic memory is typically defined as conscious memory. Yet, could it be that these birds also have a conscious recollection of their caching activities? Their flexible behaviour may be suggestive of diachronic experience.

Birch et al. (2020) draw on several studies to suggest that corvids have diachronic experience (pp. 794–795), some of which are worth mentioning here. For instance, Birch et al. mention that scrub jays have been shown to be able to remember the what, when, and where of their caching activities (Clayton et al., 2001). Furthermore, Birch et al. also note that corvids generally show a great deal of flexibility in the temporal structure of patterns that can be learnt, such as that some food will be inedible within a certain time frame, but will be sufficiently "ripe" after enough time has passed, a form of retrospective cognition and learning (de Kort et al., 2005). Recall from Chapter 2 the paradigm of future planning, where Birch et al. mention the research of Cheke and Clayton (2012) in which Eurasian jays (*Garrulus glandarius*) have been shown to be able to overcome their current desires in anticipation of future needs and to strategize accordingly, a trait that occurs relatively late in human development. One of the most astonishing findings Birch et al. (2020) mention in this context comes from Kabadayi and Osvath (2017), who demonstrated that ravens (*Corvus corax*) can anticipate how useful tokens or tools can be for future tasks, even when they are entirely novel, thus rivalling the tool-use and bartering capabilities of the great apes.

It thus appears plausible that episodic memory plays an important adaptive role for corvids, enabling them to utilize their intelligence to plan for the future and make use of past memories. What these findings of spontaneous, flexible, and complex planning and anticipation of future needs and desires in response to new information provide are "promising nonverbal indicator[s] of conscious temporal integration" (Birch et al., 2020, p. 796). For an excellent review of mental time travel in corvids, consider Cheke and Clayton (2010). Can these results challenge our human-centric thinking about diachronic unity?

Like Ginsburg and Jablonka (2019), I think that this form of mental time travel constitutes the "evolution of a new type of consciousness that builds on, but goes beyond, the minimal consciousness that we have described up to this point", moving us to what Dennett described as *Popperian creatures* capable of imagination (p. 442). Studies such as these should force us to take seriously the possibility that some corvids might even have richer diachronic experiences than humans, though perhaps not all humans. When it comes to consciousness, there is much more diversity in nature than those working in the science of consciousness would typically like to admit. Those like myself and other philosophers of mind such as Colin Allen[10] suffering from *aphantasia* will surely appreciate this point. We have a reduced, if not absent, ability to engage in the voluntary conjuring of mental images and in sensory-rich mental time travel, while nevertheless having access to semantic memory, i.e. memory of facts, including those about our own past experiences (Keogh & Pearson, 2018; Pearson, 2019; Zeman et al., 2015). We cannot engage in the reliving or simulation of mental images, smells, touch, and sounds. Yet, we are cognitively capable enough to at least do philosophy. Since these capacities are incredibly complex and play such a central role in the experience of humans, one would expect that poorer richness in imagination and mental time travel would lead to losses in some functional capacities. Yet, researchers of the phenomenon were long troubled by the difficulty of finding tests for it, precisely because aphantasics appear to be able to engage in all normal human tasks. The difficulty in detecting aphantasia in human populations may provide good evidence that a high degree of diachronic unity is not necessary for consciousness to be functional.

It should be admitted, however, that research into aphantasia is still in its infancy and we may find better tests in the future to probe the functional capacities of this strange phenomenon, which may well turn out to be pathological. Recent experimental work by Bainbridge et al. (2021) provides compelling evidence for some impaired functions of aphantasics, through drawing tests relying on memory recall. Aphantasics show "impaired object memory, but intact verbal and spatial memory during recall of real-world scene images. Collectively, these results suggest a dissociation in object and spatial information in visual memory" (Bainbridge et al., 2021, p. 171). Such results are interesting, but there is also some evidence for potential benefits of aphantasia such as "higher memory precision than control participants on some measures, including significantly fewer memory errors and fewer editing in their drawings" (Bainbridge et al., 2021, p. 170). Perhaps this could be related to the memory of facts, which may be easier to correctly remember than experienced events, which are often misreported. Due to a current lack of knowledge about aphantasia, such hypotheses remain speculative, but in the context of evolution it may be tempting to think that improved conceptual and symbolic thinking has enabled humans to become less reliant on conscious memory and thinking. We have truly transcended a stage where we are just Benthamite creatures.

Could corvids similarly rely on semantic memory (see also Clayton et al., 2003)? Ginsburg and Jablonka (2019) appear to think so, noting that "it may well be that the recall is only richly semantic (the jay has semantic knowledge of where and when and what it cached but does not relive the past experience)" (p. 445). On the contrary, I expect that a certain richness in semantic memory may require higher neural complexity than "mere" episodic recall. Whereas a human who cannot feel pain will face difficult problems in their lives, they can counteract it in a cognitive manner that would not be available to most animals with that condition – instead, they would die. This is why I suspect that consciousness in the sense of felt experiences, rather than rich conceptual thought, is more important to many animals than it is to humans. Instead of thinking about the consciousness profiles depicted in Chapter 2 as having humans covering the richest kinds of each dimension, it may very well be possible that other animals outrank us in more than just smell and touch. And if someone with aphantasia can study mental time travel and visual imagination in other humans, without themselves having this experience, there is no reason to think that we cannot investigate subjective experiences in other animals that may be widely different from what we are familiar with, such as echolocation in bats and infrared sensing in snakes.

Finally, we have to be careful not to be misled by our human conscious experience, an experience that is typically highly integrated and includes episodic memory. Using introspection can tempt one to confuse insights into one's own subjective experience with laws of necessity about experience itself. Perhaps I am "lucky" here since I fare quite poorly on at least one dimension of the temporal side of things. Philosophy of mind would be much enriched if it recognized that diversity in subjective experience within our species is as much the norm as for any other trait. By mapping out the intra-species variations and gradations of human experience, we would make great progress towards recognizing inter-species variations and gradations of consciousness as a phenomenon in nature. The more we learn about this elusive phenomenon and its role in nature, the harder it is to maintain that it is – as Griffin put it – a typologically discrete, all-or-nothing quality unique to a species emerging roughly 300,000 years ago in the African savanna.

5.7 Challenges, Conclusion, and Further Directions

In this penultimate chapter, I had the goal of bringing evolutionary and ecological thinking to the science of consciousness and showing that the other dimensions of consciousness can readily be seen as "add-ons" – or rather, enrichments – of an evaluative core. This led me to discuss numerous animals such as snails, slugs, fruit flies, crabs, bees, octopuses, fishes, and reptiles of both the avian and non-avian sorts. The goal was to demonstrate that we need an ecological and ethological understanding of these animals to develop a truly biological science of consciousness, one that begins with a true understanding of what makes a bat a bat, a snake a snake, and a healthy bee a healthy bee. While this work may appear speculative, we have to seek an understanding of the species-specific pathological complexity of their life histories, even if many will suspect them to lack sufficient nervous system complexity to warrant an attribution of consciousness.

As I hope to have made clear throughout this book, it is precisely with such simple animals that we have to begin an evolutionary investigation of consciousness as a phenomenon that gradually came into existence. The animals usually seen as belonging to a "grey-zone" are the best cues to what the gradual evolution of subjective experience may have been like, providing us with a rich diversity of different life-history strategies. My discussion of gastropods and arthropods was framed against recent work by Godfrey-Smith (2020b, 2020c), who emphasized the possibility of a disassociation between the sensory and evaluative side in gastropods and insects, which may yield us a two by two table representing capacities across each dimension. Whereas most discussions of consciousness focus on animals that plausibly have together at least a minimal sense of both sensory and evaluative experience, such as most vertebrates and octopuses, a comparative bottom-up approach allows us to reverse-engineer the origins of consciousness by emphasizing the animals in which consciousness exists in only a rudimentary form and possibly in a limited subset of dimensions.

Furthermore, by looking at non-human animals, we have been able to challenge assumptions about the other dimensions of consciousness and think about their potential adaptive roles within the life histories of different animals. In Figure 5.1, I have offered suggestive comparative consciousness profiles based on the discussions in this chapter. The numbers here are not meant to represent a scale or ordering in a strong sense – only a useful visualization to think about the experiences of these animals. If consciousness can be broken down into different dimensions, we ought to expect a great diversity in the phenomenological complexity of different animals, dependent on their pathological complexity. My goal here was only to walk some of the distance towards a true biological science of consciousness that takes a comparative approach to heart.

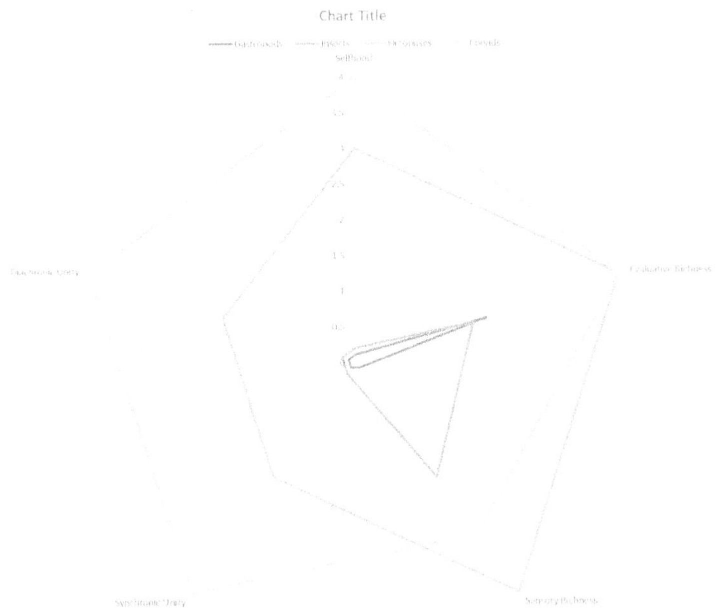

FIGURE 5.1 Hypothetical consciousness profiles for the phenomenological complexity of gastropods, insects, octopuses, and corvids [modelled after Birch et al. (2020, Figure 1, p. 791)]

As Schnell and Clayton (2021) once pointed out, thinking about animal cognition has largely focused on the "challenges experienced by large-brained vertebrates and do not accommodate taxa that have diverse life-histories or that have been exposed to different types of evolutionary pressures" (p. 29). My hope is that the pathological complexity framework can help us to achieve a broadening of scope and that others will join the cause of trying to answer the question of what it is like to be a bee, an octopus, or a raven.

Notes

1 The neuroscientist and ventriloquist Michael Graziano (2016) frequently uses a small Orangutan puppet named Kevin in his talks to demonstrate how quick we are to attribute consciousness to other entities.
2 Dennett (2019a) reports of Cog, a robot developed by a team at MIT that "moved its arms and eyes and head with such humanoid vivacity and even grace that naive observers often blurted out loud their startled conviction that it was conscious" (p. 2).
3 See Walters (2018) for a recent review.
4 See Caporael et al. (2014), Griesemer (2014), and Veit (2022d) for discussions of "scaffolding" in evolution.
5 See also Wigglesworth (1980).
6 In personal communication, Godfrey-Smith admitted that he should not have used the term "grooming" in his list.
7 Crustaceans are likely a paraphyletic group (Blackstone, 2001).
8 Eisemann et al. (1984) cited the work of the influential British octopus researcher Martin Wells (1978) as a potential source for making a case for octopus pain.
9 Especially in eutherian mammals, who have even more brain connectivity between both halves (see Godfrey-Smith, 2021a).
10 Personal communication during my time working as a pre-doc at CMU in 2019.

6

THE FINAL, CROWNING CHAPTER OF THE DARWINIAN REVOLUTION

6.1 Introduction

We are nearing the end of our journey. Since we have covered a lot of ground thus far, I will conclude this book by providing a succinct summary of the core arguments presented, evaluating how far we've come in addressing the goals I set out in the beginning, and offering an optimistic outlook for future investigations.

6.1.1 Chapter Outline

This chapter is structured as follows. In Section 6.2 "Summary", I offer a brief recap of the foregoing chapters. In Section 6.3 "Consciousness Darwinized", I synthesize these chapters to show how far we have come in making sense of consciousness within a biological view of the mind. Lastly, Section 6.4 "Final Thoughts and Future Directions" offers tentative suggestions for future research on animal consciousness and the pathological complexity thesis.

6.2 Summary

This monograph opened with an extended preface to offer a gentle introduction to and motivation for the naturalist project undertaken within. This project aimed to make sense of the place of mind in nature, both in the sense of understanding other minds and the general relationship between mind and matter – two problems to which answers now vary so widely that one may legitimately think that consciousness is as much, if not more, of a mystery as ever. As I hope to have made clear, however, it is within a *Darwinian standard* that the study of consciousness is provided with a much needed theoretical bottleneck that would constrain and aid the development of a truly biological science of consciousness. This is why, at the beginning of Chapter 1, I offered a quote by Griffin in which he maintained that the evolutionary study of consciousness may constitute "the final, crowning chapter of the Darwinian revolution" (1998, p. 14).

6.2.1 Ch. 1: A Darwinian Philosophy for the Science of Consciousness

Following Griffin's call for a cognitive ethology of consciousness, Chapter 1 aimed to make the case that any biological approach to consciousness must address the *teleonomic question* of what

DOI: 10.4324/9781003321729-6

consciousness in all of its gradations and varieties does for healthy agents within their normal ecological lifestyles and the natural environments in which they have evolved. To do this, I argued that an evolutionary bottom-up approach to consciousness is both possible and necessary to advance a true Darwinian science of consciousness. But to advance such an approach, I also offered my own articulation of a thesis on the function of consciousness – the pathological complexity thesis.

Pathological Complexity Thesis:
The function of consciousness is to enable the agent to respond to pathological complexity.

Part of my motivation for introducing the pathological complexity thesis was to establish a close connection between the natural phenomena of health and consciousness. Firstly, the pathological complexity thesis elaborated the idea that the origin and function of consciousness lie in enabling organisms to efficiently deal with their species-specific pathological complexity, i.e. the trade-offs of their life histories. To understand health as a natural phenomenon is simply to measure how well an organism succeeds at dealing with their pathological complexity trade-offs, with fitness providing the common currency for biological design. And the origins of consciousness can be understood as the evolution of hedonic valence as a proximate common currency to enable organisms to efficiently calculate these trade-offs in their own actions.

But this was not the only link between health and consciousness that I wanted to emphasize. Drawing on the history of the Darwinian revolution and its extension towards behaviour, I also argued that an extension of this revolution towards consciousness must begin with an appreciation for the distinction between healthy and pathological variations of consciousness. For this, however, we are in need of both a comparative study of consciousness across the animal tree of life and a naturalist understanding of health, which I argued can be derived from our current best theory of organisms. Such a theory will in turn enable us to understand them as active *agents* and *subjects*, including their subjective experience as an integral part of our biological understanding of what makes a bat a bat, a snake a snake, and a healthy bee a healthy bee.

Making use of a number of lessons from the Darwinian revolution, the chapter concluded with an introduction to the state-based behavioural and life-history theory as the teleonomic theory of organismal agency and thus the ideal tool with which to operationalize my concept of pathological complexity. As I hoped to have made clear, it is within this theory that we can bring out the subject-side of organisms and satisfy Lewontin's demand to bring Darwinism to completion, by paying attention to the "functional needs" of the organism (1985, p. 85). Lastly, while important names in the field, such as Godfrey-Smith (2020b), repeatedly speak of a particular animal lifestyle emerging in the Cambrian that led to the evolution of consciousness, this idea has thus far been left rather vague. As a theory of the diversity of "animal lifestyles", the life-history theory is the obvious resource with which to explicate this idea.

6.2.2 Ch. 2: The Explanandum: Animal Consciousness and Phenomenological Complexity

Following my emphasis on the healthy and pathological varieties of consciousness, Chapter 2 argued that any biological theory of consciousness must account for the gradations and varieties of consciousness both within and across species, i.e. what I have dubbed *phenomenological complexity*, as the explanandum of this book.

In order to show that a breakdown of consciousness into multiple dimensions, rather than as an all-or-nothing phenomenon, is not only possible but can also be in principle measured in non-human animals, I have reviewed and discussed the call by Birch et al. (2020) to develop multi-dimensional frameworks for animal consciousness, including their own distinction between the five dimensions of sensory experience, unity, temporality, selfhood, and lastly evaluative experience.

By discussing their suggested experimental paradigms to assess these dimensions in animals, in addition to some further ones I introduced, I hope to have removed (or at least weakened) any initial reservations to the possibility of advancing a Darwinian science of consciousness that removes humans from the centre of reference, as was the case with the Darwinian revolution in physiology and behaviour. Both these dimensions and the experimental paradigms associated with them went on to play major roles in the following chapters.

6.2.3 Ch. 3: The Origins of Consciousness or the War of the Five Dimensions

Making use of the distinctions between the five dimensions, Chapter 3 attempted to tease apart the problem of consciousness by treating it like an onion, peeling away one of these dimensions after another until only one remained: the most plausible contender for the dawn of subjectivity. Rather than treating consciousness as a complex all-or-nothing phenomenon that makes a gradualist evolution seem impossible, I took an evolutionary reverse-engineering approach that tried to narrow the explanatory gap by re-conceiving consciousness as a multidimensional phenomenon, in which each dimension could have been built upon one other.

Firstly, I argued that the two dimensions of unity should be seen as ways that experience can be organized, not elements that are necessary for the existence of qualitative experience. The origins of qualia are likely to have been much more akin to fragmented experiences that only later come to be integrated into coherent points of view.

Secondly, I argued that the two dominant traditions in the philosophy of mind and the science of *human* consciousness – strongly externalist representationalist theories of consciousness that overemphasize sensory experience and strongly internalist ones that overemphasize self-awareness– have forced us into a false dilemma, with each way of thinking about consciousness seemingly neglecting the other.

While we cannot rule out just yet theories of consciousness built on these dimensions, I argued that the dimension of evaluation can offer us a better dynamic model of consciousness, centred on feedback – one that keeps what is good about both approaches, but discards the rest. Furthermore, it is this dimension that appears to fit better with how laypeople think about consciousness as hedonic valence and to minimize the hard problem challenge, since it is precisely the felt aspects of this capacity that make it functional. And with this rearranged focus on minimal consciousness, the explanatory gap has been significantly reduced.

Unfortunately, Tye (2021) was right to note that "[t]he general topic of the origins of qualia is not one on which philosophers have said a great deal". The goal of my next chapter was thus to address this omission.

6.2.4 Ch. 4: Pathological Complexity and the Dawn of Subjectivity

My goal in Chapter 4 was to offer an explanation for the evolutionary origins of consciousness. It was here that I explicated the pathological complexity thesis in detail.

Firstly, I defended my emphasis on pathological complexity as *the* teleonomic measure of biological complexity that we should focus on – rather than, say, environmental complexity, nervous system complexity, or any other kind of complexity measure we may come up with.

Secondly, I argued that the evolution of hedonic valence can be understood as an adaptive response to a computational explosion of pathological complexity that occurred during the

Cambrian, allowing Darwinian agents to track what matters to them and to make the right decisions at the right time. It is in this context that proximate interests or imperative motivations are born, giving rise to a new mode of animal agency in these *Benthamite creatures* with efficient action control and action selection.

Lastly, I argued that sensory experience plausibly evolved as an enrichment in the discriminatory capacities of evaluation, whereas minimal selfhood plausibly evolved in further enrichments on the sensory side, involving the distinction between interoception and exteroception. Integration of experience at a time and across time can simply be seen as further transformations in the structure of consciousness, instead of constituting their own explanatory gaps. It is ultimately their being part of an evaluative system that makes these capacities felt.

6.2.5 Ch. 5: Pathological Complexity Meets Phenomenological Complexity

In Chapter 5, the pathological complexity thesis was put to the test by responding to an argument by Godfrey-Smith that there could be a phylogenetic split between animals who plausibly only have evaluative experience (i.e. gastropods) and those who only have sensory experience (i.e. insects). Since the pathological complexity thesis rests on the idea that the core and origins of consciousness lie in hedonic evaluation, I used the former group as evidence for my thesis, whereas the latter constituted a challenge that had to be overcome.

By reviewing the literature and discussing the life histories of arthropods, I argued that even those animals that are presently the strongest case for the independent existence of sensory experience turn out to have rich evaluative capacities after all. Furthermore, I discussed octopuses as a challenge to our human-centric thinking about selfhood, fish and non-avian reptiles with their natural split brains as a challenge to synchronic unity, and finally corvids with their spectacular memory capacities as a challenge to our thinking about diachronic experience. In line with the mission of this book, these discussions also involved recourse to pathological conditions of consciousness, such as aphantasia.

Finally, the chapter was meant to demonstrate the usefulness of my evolutionary pathological complexity framework by putting the life histories of animals centre stage in theorizing about animal consciousness. By linking pathological complexity to phenomenological complexity, we are offered a bidirectional approach that can make predictions and feed back into our understanding of the lives of animals, just as the cognitive ethologists originally intended.

6.3 Consciousness Darwinized

While each of the chapters made their own distinctive contributions, and hence could to some degree be read independently, this book has largely been one long argument to defend a single thesis and framework. In order to realize Griffin's vision of a Darwinian science of consciousness, I argued that we must free ourselves from perhaps the most pernicious pre-Darwinian dogma that has held us back for far too long: the faith and confidence in the uniqueness and superiority of the human mind (Griffin, 1981, p. 163). For whatever negative impact their tradition still has on the study of animals and consciousness, it was the great insight of the behaviourists of the twentieth century that a psychology that makes consciousness *as we humans experience it* the reference point of all our theories of behaviour would force us into the same situation that had been faced by biology and medicine prior to the Darwinian revolution. Yet, despite the return of a so-called science of consciousness, the field still suffers from just the same flaw as did its precursor in the nineteenth century: a failure to recognize evolutionary continuity. A central motivation of this book has been to remove these shackles of a pre-Darwinian view of nature and do for consciousness what the behaviourists have done for behaviour: *to remove humans from our centre of reference.* Instead of accepting this important insight, modern consciousness science operates under the conditions of highly constrained experiments, allowing few degrees of freedom and studying a only small number of animals (mostly humans).

The pressure to make the science of consciousness as objective as possible has forced it into a dilemma between externalist and internalist approaches that resembles the pre-Darwinian study of life and behaviour. The problem with a lack of thinking about ecological feedback is that we are missing out on the core of the Darwinian revolution. The absence of ecological thinking in highly constrained experiments is sometimes framed as a conflict between internal validity and ecological validity (i.e. the credibility that our experiment has no bias versus the applicability of our experimental results to the real world), but this obscures what is at stake. In the study of a complex phenomenon, if we frame the problem in this manner, then it will always appear to be the rational move to begin with highly constrained situations, but this dictum must be taken with a grain of salt when it comes to Darwinian phenomena. As I have argued in Chapter 1, the Darwinian revolution did not begin with the study of animals in the lab – rather it came out of natural history, i.e. the observation of healthy and pathological cases of living activity in a natural setting. This is why the ethologists maintained that an extension of the Darwinian revolution to behaviour must begin here: with an understanding of the natural healthy lives of animals.

We need to ground our scientific theorizing firmly within an ecological understanding of the organism. To demand experimental paradigms occurring prior to this stage to fit within the straightjacket of highly controlled experiments, such as those conducted in physics, would have severely held back the modern biology of behaviour. But whereas the research areas of animal behaviour and cognition now have a healthy pluralism of accepted methods and experiments – combining fieldwork, ecological models, and constrained experiments in the lab – the study of consciousness still suffers from the old conflict between "comparative" psychology and the truly comparative methods of ethology. The demand for scientific "objectivity" has put consciousness science into this straightjacket of highly controlled experiments, something that is doubly problematic if, as I have argued here, consciousness evolved to deal with the pathological complexity caused by high degrees of freedom. This is the important lesson we should take away from the ethologists' Darwinian revolution in studying behaviour: in order to study a biological phenomenon, we must use a truly comparative approach that attends to the similarities and differences in octopuses, bees, bats, insects, and the like by investigating what it means to be a healthy agent of that species acting in the normal ecological setting in which it evolved. And such a Darwinian approach must also recognize a multiplicity of gradations and variations in the subjective experience both within and across different species, which I have called phenomenological complexity.

To begin the final, crowning chapter of the Darwinian revolution must mean to learn from the Darwinian revolution of life and to force the problems of consciousness through this theoretical bottleneck of evolutionary theory. This is the standard that the study of consciousness was so desperately lacking. As I have hoped to show in this book, modern evolutionary biology – and in particular state-based life-history theory – offers us rich theoretical resources and constraints within which to think about the problems of mind. At the forefront of a biological science of consciousness must be the functionalist question of what consciousness in all of its diversity and gradations does *for* healthy sentient agents within their normal ecological lifestyles and the natural environments in which they have evolved. And in order to answer this question, we must attend to the very origins of this phenomenon, rather than its special presentation in humans. A truly Darwinian comparative approach cannot make the human case the model for all consciousness.

Such a science of consciousness will inevitably transform our understanding of the phenomenon, similarly to how the science of space and time has done so for their target phenomena.[1] Philosophers such as Nagel are sceptical of this idea, since they think that our understanding of consciousness must be anchored in our first-person experience. But a true biological science of consciousness needs only to honour our experience as one variety of consciousness, a special case of a much more widespread and diverse phenomenon. Just as theories of space-time should account for how space and time appear for humans as a special case of a more general phenomenon, so can a science of consciousness transform our understanding of this phenomenon as one that is complex and multidimensional, with plenty of varieties and gradations of phenomenological complexity. As Bryce Huebner and I have argued in a commentary in *Animal Sentience*, the "future of animal sentience research lies not in drawing boundaries but in empirically investigating *what it feels like to be* an echo-locating bat, an infrared-sensing snake, an octopus with multiple distributed ganglia, a fish without a neocortex, or an arthropod such as a spider or a honey bee" (Veit & Huebner, 2020, p. 3).

Not only have I argued that we must begin with a teleonomic understanding of the organism, i.e. what organisms are *for*, which will enable us to make distinctions between the normal and the pathological, but I have also argued that consciousness itself is an adaptive response to complex health challenges arising at a multicellular level from an active animal lifestyle, which constitutes a major transition in the agency. Health, I argued, can be understood as a measure of how well an organism deals with its pathological complexity, which I have argued in turn can be explicated in terms of modern state-based behavioural and life-history theory. Pathological complexity corresponds to the decision-theoretic fitness maximization problem of life-history agents trying to secure their representation in the next generation. Fitness provides us with a common currency for comparing the health status of different animals, whether they have lesions, parasites, broken bones, diseases, or poor environments. Health and pathology, as Lorenz argued early on, must be understood in an ecological context. And it is in the context of the Cambrian explosion and the evolution of behavioural flexibility – of predation, fluctuating environments, and more complex lifestyles – that pathological complexity exploded, making the proximate common currency of valence a requirement in order for organisms to efficiently map their behaviour to environmental states, depending on their own bodily state. The origins of consciousness, as speculated by Darwin's protege Romanes and many other evolutionists, lie in the dawn of affective feelings, of the evaluation of states and behaviours as good, bad, or neutral.

While the available evidence cannot rule out competing theories of consciousness just yet, my theory has two major points in its favour, giving it an edge over many other contenders in the field. Firstly, the "hard problem" of why things feel a certain way does not appear to be *as much* of a challenge within a hedonic framework, as opposed to a strongly externalist representationalist one. Things feel a certain way, because they *have to* feel that way to be functional, which also provides a superior picture to strongly internalist views that make consciousness an almost automatic feature. A focus on evaluation is able to overcome the alleged inability of functionalist theories of consciousness to explain why consciousness *feels* like anything at all (see also Solms, 2021). Secondly, it is within such a model of evaluation as the basis of consciousness that the other dimensions of consciousness can readily be explained as later transformations. As I have shown in Chapter 5, the pathological complexity thesis allows us to make predictions regarding the subjective experiences of other animals such as insects, octopuses, and corvids – predictions that can be tested with the experimental paradigms discussed in Chapter 2. Phenomenological complexity can be explained as an adaptive response to pathological complexity. This is a second major advantage of my framework against most competitors that have a hard time making testable predictions, such as the integrated information theory of consciousness. And the empirical implications of the pathological complexity thesis appear to at least provide us with something like a good initial test for its viability.

Further empirical tests will be able to push it further, allowing for the integration of subjects and *subjectivity* into the Darwinian revolution, thus moving us towards Griffin's goal of creating a cognitive ethology that makes sense of the place of mind in nature through recourse to a sound evolutionary understanding of an animal's evolved life-history strategies.

6.4 Final Thoughts and Future Directions

Because of scepticism against the very idea of animal minds, researchers were long forced to make their experiments as empirically rigorous and controlled as possible, at the cost of telling us little about what the mind does *for* animals in the wild. As I noted above, this omission can be attributed to a very old and still entrenched difference between comparative psychologists and cognitive ethologists that maps onto an even older conflict between behaviourists and classical ethologists. While both groups can be interested in the evolution of traits, those in the tradition of comparative cognition are faced with a "constant pressure towards laboratory experimentation in conditions that are often of questionable ecological validity" (Allen, 2004, p. 605) and largely focus on a small range of species. Cognitive ethologists, on the other hand, emphasize the importance of observing a broad range of taxonomic groups in conditions as close as possible to the ecological conditions in which they have evolved (see also Allen, 2006). Naturally, such an approach makes experiments hard, if not impossible, in many cases. So there are inevitable trade-offs here. Nevertheless, ecological field experiments and observations can be useful in thinking about the possible adaptive value of consciousness in animals. But this could hardly be further from anything like a true test for the presence of consciousness, a demand that I hope we will be able to overcome in the near future. As Griffin put it:

> I am confident that with patience and critical investigation we can begin to discern what life is like, subjectively, to particular animals under specific conditions [...]. Cognitive ethologists can certainly improve greatly on these preliminary inferences, once the creative ingenuity of scientists is directed constructively toward the important goal of answering Nagel's basic question: What is it like to be a bat, or any other animal?
>
> *Donald R. Griffin (1992, p. 260)*

While the dividing lines between behaviourists and ethologists have weakened, they are still very much alive when it comes to the study of animal consciousness, and the two will have to become integrated for a true biological science of consciousness to emerge. As Allen and Bekoff (1997) have argued early on, we are in need of *species-fair* tests that are adapted to the different lifestyles of different taxa (p. xiv). Tests for diverse species must include a sound evolutionary and ecological understanding of their lives in order to avoid premature conclusions about what they can or cannot experience, and I hope that the pathological complexity thesis can help us in thinking about the design of such tests.

The evolutionary account I have tried to sketch here necessarily had gaps. In doing philosophy of nature, one cannot be concerned with the precise nor the causal specifics of any particular situation. Instead, one draws on and synthesizes evidence from a variety of different disciplines to make sense of the world on a very general level – in our case the place of mind in nature, both in the sense of the connection between matter and mind and the literal locations of conscious agents in the world. In trying to develop a general framework for this task – a model for thinking about consciousness in a bottom-up fashion across the tree of life – there are inevitable theoretical trade-offs. But that is just what has to be provided at the beginnings of a true biological science of a complex phenomenon. The question is not how a particular bat is conscious, but how consciousness itself can be made sense of in the most general form as a phenomenon in nature. This general approach is not anti-neural as much as it is functionalist. We are asking for the teleonomic reason for the existence of subjective experience.

Future work will inevitably improve upon the framework I have tried to provide here. But none of this will be a problem if it can serve as a theoretical scaffold for future bottom-up approaches that take Darwinian thinking seriously, emphasizing both gradations and varieties of subjective experience in animal life. What has been shown, I hope, is that this framework is not just a "just-so story". All living agents have to deal with the complex design problem of trading off different functional demands against each other. Just as fitness constitutes an ultimate common currency for evaluating the health status of an organism, so does hedonic valence provide a proximate common currency for agents to deal with a degree of pathological complexity that would otherwise become unmanageable for a behaviourally flexible, vulnerable, multicellular organism. Consciousness evolved to enable animals to efficiently deal with this pathological complexity by providing a dynamic way of evaluating trade-offs in the deployment of alternative actions.

One curious feature of this account is that this explosion of pathological complexity may in principle have occurred in separate lineages, so just as there are several independent origins of behavioural complexity, there could be multiple independent origins of subjective experience. Godfrey-Smith (2020b) does not attempt to offer a conclusive answer but raises the possibility that consciousness may have independently arisen in cephalopods, arthropods, and vertebrates. Some locate consciousness only in vertebrates or mammals, because of their similarities to us, but this view is making less sense the more we learn about other branches of life. The lifestyles of cephalopods and crustaceans are too similar to those of many vertebrates to deny the adaptive value of sentience in these creatures, and there is now plenty of experimental evidence that supports the attribution of pain to them (Birch et al., 2021). Whether this common evaluative core arose once or multiple times is an open question, but the model presented here can make sense of an early occurrence of this capacity as a plausible candidate scenario. As a much more basic capacity is required to even allow for such behavioural flexibility to become manageable for the organism, it would explain the difference between Avalon and Cambrian explosions.

However, while I have offered an alternative to views associating consciousness with later evolutionary innovations, such as rich forms of information integration or selfhood, this doesn't mean proponents of such views must be entirely wrong. Indeed, I see it as a virtue of my account that their claims regarding a more recent dawn of consciousness may very well be re-conceived of instead as what Godfrey-Smith (2016c, 2016d) has described as a *transformation view* as opposed to a *latecomer view*. A transformation view accepts that "some late-evolving features of our brains do greatly *affect* the nature of subjective experience" but does not treat them as bringing consciousness into existence (Godfrey-Smith, 2016c, p. 499). The transformation view alongside the pathological complexity thesis makes the problem of minimal consciousness significantly smaller and narrows the explanatory gap through the use of the hedonic picture, but it also has the virtues of accommodating the evidence presented by latecomer views. While latecomer views can give the impression of being less speculative by asking for a great degree of similarity between other animals and us in order to grant them consciousness, they commit the fallacy of failing to distinguish human consciousness from consciousness as a phenomenon in nature, which plausibly has a much simpler origin.

Indeed, if we want to hold onto both the idea that consciousness has only arisen once and that it is shared by these three lineages, it must have arisen in a common ancestor that was very simple. And as far as I have argued here, the pathological complexity thesis makes more sense of such an early occurrence than any of its competitors (such as those accounts provided by Godfrey-Smith or Ginsburg and Jablonka) that emphasize later features arising independently in these three lineages. Somewhere around the beginning of the Cambrian explosion, if not earlier, lived the last common ancestor of arthropods and vertebrates. It is plausibly here, in some worm-like creature resembling modern annelids, that we find something like the very first sparks of feelings. However, I do not want to assert that all descendants of these semi-proto-quasi sentient creatures must be conscious. As I noted in my discussion of insects, some branches of animal life could have become simpler and "lost" sentience. After all, as I have argued throughout this book, consciousness must ultimately pay off for organisms, and simpler lives with lower pathological complexity may well succeed without hedonic valence.

The pathological complexity thesis was ultimately intended as a teleonomic theory of consciousness, but it may also offer us some clues regarding the metaphysics of consciousness. In particular, we should plausibly think of consciousness in terms of something like a hybrid materialist-functionalist identity thesis that identifies the origin of consciousness with the origin of valence systems in multicellular animals, somewhere around the beginning of the Cambrian. To be such an animal simply means to be a subject with experience. Tracing the gradual evolution of Benthamite creatures from objects into subjects/agents makes the mystery of consciousness substantially smaller than it would otherwise seem.

Perhaps the most important consequence of the ideas in this book is the relevance of the connections between pathological complexity and sentience for the purposes of ethics and policy-making, since it is sentience that is usually taken to make an entity a subject of moral concern (Browning, 2020c; Browning & Veit, 2022c). Some dub this view "sentientism" – the view that "you have moral status, i.e. you are a subject of moral concern, if and only if you are sentient, i.e. if and only if you are capable of phenomenally consciously experiencing pleasure or pain" (Sebo, 2018, p. 4). As my interest in animal consciousness was also motivated by ethical concerns, I have published several papers on animal ethics, sentience, and welfare science in collaboration with Heather Browning,[2] but one problem we were repeatedly faced with is the challenge of inter-species comparisons of welfare: i.e., the more broadly we attribute sentience to other animals, the less reasonable it will be to assign equal moral weight to all insects, birds, and octopuses (see Browning, 2023 for a detailed examination). Their capacity to suffer and experience pleasure reasonably scales according to their degree of consciousness, which forces us to think about consciousness in a gradualist, rather than "on" or "off" manner. While we have offered some brief discussions on animal sentience in relation to the life histories of different animals (Browning & Veit, 2021c), the pathological complexity thesis may offer us a useful evolutionary proxy measure to assess different levels of evaluative richness in the subjective experience of different animals. How to measure animal welfare is a notoriously difficult problem (Browning, 2022), but a measure of pathological complexity would enable us (at least in principle) to assess animals according to a so-called sentience multiplier in calculations for how to best improve animal welfare (see Browning, 2020a), and this could in turn be tested with the experimental paradigms for evaluative richness discussed in Chapter 2. But this will be a task for future work.

A similar path may be opened for discussions of consciousness in robots and artificial intelligence, the presence of which I believe would require conditions and abilities approximating the pathological complexity of animal lifestyles – something that has not yet been achieved. It may well be that the particular material basis of evaluation in animal agents matters for consciousness, but the functionalist approach offered here could at least offer us clues for how to build sentient machines. To understand consciousness, we will first have to understand its evolutionary origins, rather than speculate about consciousness in non-biological entities. Other future applications of the framework may also extend to the explanation of disorders and varieties of consciousness in humans. Since my being on both the autism and aphantasia spectrums has informed my awareness of the phenomenological diversity of minds, I hope to expand into this area in the future.[3]

Finally, I hope there is some truth in the account I have offered and that my methodology of reverse-engineering the origins of consciousness within the hedonic evaluation and the pathological complexity of animal life has moved us a few steps closer towards the writing of the final, crowning chapter of the Darwinian revolution and the development of a true biological science of animal consciousness.

Notes

1 I thank Paul Griffiths for this suggested analogy.
2 See Veit and Browning (2020a, 2021) and Browning and Veit (2020a, 2020c, 2021b, 2022a).
3 I already have a few publications on the topic of autism (Browning & Veit, 2022d; Chapman & Veit, 2020, 2021), but I hope to apply my framework in the future to the evolution of mindreading abilities.

BIBLIOGRAPHY

Adam, R., & Güntürkün, O. (2009). When one hemisphere takes control: Metacontrol in pigeons (Columba livia). *PLoS One, 4*(4), e5307.

Adamo, S. A. (2016). Do insects feel pain? A question at the intersection of animal behaviour, philosophy and robotics. *Animal Behaviour, 118*, 75–79.

Albrecht, G. L., & Devlieger, P. J. (1999). The disability paradox: High quality of life against all odds. *Social Science & Medicine, 48*(8), 977–988.

Allen, C. (2004). Is anyone a cognitive ethologist? *Biology & Philosophy, 19*(4), 589–607.

Allen, C. (2006). Transitive inference in animals: Reasoning or conditioned associations. In S. Hurley & M. Nudds (Eds.), *Rational animals?* (pp. 175–185). Oxford University Press.

Allen, C. (2013). Fish cognition and consciousness. *Journal of Agricultural and Environmental Ethics, 26*(1), 25–39.

Allen, C. (2017). Associative learning. In K. Andrews & J. Beck (Eds.), *The Routledge handbook of animals minds* (pp. 401–408). Routledge.

Allen, C., & Bekoff, M. (1997). *Species of mind: The philosophy and biology of cognitive ethology*. MIT Press.

Allen, C., Fuchs, P. N., Shriver, A., & Wilson, H. D. (2005). Deciphering animal pain. In M. Aydede (Ed.), *Pain: New essays on Its nature and the methodology of Its study* (pp. 351–366). MIT Press.

Allen, C., & Trestman, M. (2017). Animal consciousness. In E. N. Zalta (Ed.), *The Stanford encyclopedia of philosophy* (Winter 2017 ed.). Retrieved May 22, 2022, https://plato.stanford.edu/archives/win2017/entries/consciousness-animal/

Alupay, J. S., Hadjisolomou, S. P., & Crook, R. J. (2014). Arm injury produces long-term behavioral and neural hypersensitivity in octopus. *Neuroscience Letters, 558*, 137–142.

Anderson, J. R. (1986). Mirror-mediated finding of hidden food by monkeys (Macaca tonkeana and M. fascicularis). *Journal of Comparative Psychology, 100*(3), 237–242.

Anderson, J. R., & Gallup, G. G. (2015). Mirror self-recognition: A review and critique of attempts to promote and engineer self-recognition in primates. *Primates, 56*(4), 317–326.

Animal Ethics. (2020). Establishing a research field in natural sciences: Three case studies. *Oakland: Animal Ethics*, 1–62. Retrieved March 25, 2021, https://www.animalethics.org/establishing-field-naturalsciences

Antolin, M. F. (2011). Evolution, medicine, and the Darwin family. *Evolution: Education and Outreach, 4*(4), 613–623.

Appel, M., & Elwood, R. W. (2009). Motivational trade-offs and potential pain experience in hermit crabs. *Applied Animal Behaviour Science, 119*(1–2), 120–124.

Archer, J. (1992). *Ethology and human development*. Harvester Wheatsheaf.

Aristotle. (1991). On the soul (J. A.Smith Trans.). In J.Barnes (Ed.), *The complete works of Aristotle: The revised Oxford translation* (Vol. 1, pp. 641–692). Princeton University Press.

Aveling, E. B. (1886). *Die Darwin'sche Theorie*. Verlag von J. H. W. Dietz.
Avramides, A. (2020). Other minds. In E. N. Zalta (Ed.), *The Stanford encyclopedia of philosophy* (Winter 2020 ed.). Retrieved April 4, 2022, https://plato.stanford.edu/archives/win2020/entries/other-minds/
Baars, B. (1988). *A cognitive theory of consciousness*. Cambridge University Press.
Babcock, L. E., & Robison, R. A. (1989). Preferences of Palaeozoic predators. *Nature, 337*(6209), 695–696.
Bainbridge, W. A., Pounder, Z., Eardley, A. F., & Baker, C. I. (2021). Quantifying aphantasia through drawing: Those without visual imagery show deficits in object but not spatial memory. *Cortex, 135*, 159–172.
Balasko, M., & Cabanac, M. (1998a). Behavior of juvenile lizards (Iguana iguana) in a conflict between temperature regulation and palatable food. *Brain, Behavior and Evolution, 52*(6), 257–262.
Balasko, M., & Cabanac, M. (1998b). Motivational conflict among water need, palatability, and cold discomfort in rats. *Physiology & Behavior, 65*(1), 35–41.
Balduzzi, D., & Tononi, G. (2008). Integrated information in discrete dynamical systems: Motivation and theoretical framework. *PLoS Computational Biology, 4*(6), e1000091.
Balduzzi, D., & Tononi, G. (2009). Qualia: The geometry of integrated information. *PLoS Computational Biology, 5*(8), e1000462.
Balleine, B. W., & Dickinson, A. (1998). Goal-directed instrumental action: Contingency and incentive learning and their cortical substrates. *Neuropharmacology, 37*(4–5), 407–419.
Bard, K. A., Todd, B. K., Bernier, C., Love, J., & Leavens, D. A. (2006). Self-awareness in human and chimpanzee infants: What is measured and what is meant by the mark and mirror test? *Infancy, 9*(2), 191–219.
Barlassina, L. (2020). Beyond good and bad: Reflexive imperativism, not evaluativism, explains valence. *Thought: A Journal of Philosophy, 9*(4), 274–284.
Barlassina, L., & M. K. Hayward (2019). More of me! Less of me!: Reflexive imperativism about affective phenomenal character. *Mind, 128*(512), 1013–1044.
Baron-Cohen, S. (1997). *Mindblindness: An essay on autism and theory of mind*. MIT Press.
Baron-Cohen, S. (2000). Theory of mind and autism: A review. *International Review of Research in Mental Retardation, 23*, 169–184.
Barrett, L. F. (2017). *How emotions are made: The secret life of the brain*. Pan Macmillan.
Barron, A. B., & Klein, C. (2016). What insects can tell us about the origins of consciousness. *Proceedings of the National Academy of Sciences, 113*(18), 4900–4908.
Barron, A. B., Søvik, E., & Cornish, J. L. (2010). The roles of dopamine and related compounds in reward-seeking behavior across animal phyla. *Frontiers in Behavioral Neuroscience, 4*, 163.
Bateson, M., Desire, S., Gartside, S. E., & Wright, G. A. (2011). Agitated honeybees exhibit pessimistic cognitive biases. *Current Biology, 21*(12), 1070–1073.
Beer, C. (2020). Niko Tinbergen and questions of instinct. *Animal Behaviour, 164*, 261–265.
Bellingham, J., Morris, A. G., & Hunt, D. M. (1998). The rhodopsin gene of the cuttlefish Sepia officinalis: Sequence and spectral tuning. *Journal of Experimental Biology, 201*(15), 2299–2306.
Bengtson, S. (2002). Origins and early evolution of predation. *The Paleontological Society Papers, 8*, 289–318.
Bentham, J. (1879). *An introduction to the principles of morals and legislation*. T. Payne & Son.
Benton, M. J., & Donoghue, P. C. (2007). Paleontological evidence to date the tree of life. *Molecular Biology and Evolution, 24*(1), 26–53.
Bergson, H. (1920). *Mind-energy: Lectures and essays* (W. Carr, Trans.). Henry Holt.
Bermond, B. (2001). A neuropsychological and evolutionary approach to animal consciousness and animal suffering. *Animal Welfare, 10*(1), 47–62.
Berridge, K. C. (1996). Food reward: Brain substrates of wanting and liking. *Neuroscience & Biobehavioral Reviews, 20*(1), 1–25.
Berridge, K. C. (2009a). Wanting and liking: Observations from the neuroscience and psychology laboratory. *Inquiry, 52*(4), 378–398.
Berridge, K. C. (2009b). 'Liking' and 'wanting' food rewards: Brain substrates and roles in eating disorders. *Physiology & Behavior, 97*(5), 537–550.
Billard, P., Clayton, N. S., & Jozet-Alves, C. (2020). Cuttlefish retrieve whether they smelt or saw a previously encountered item. *Scientific Reports, 10*(1), 1–7.
Birch, J., Broom, D. M., Browning, H., Crump, A., Ginsburg, S., Halina, M., Harrison, D., Jablonka, E., Lee, A. Y., Kammerer, F., Klein, C., Lamme, V., Michel, M., Wemelsfelder, F., & Zacks, O. (2022). How should we study animal consciousness scientifically? *Journal of Consciousness Studies, 29*(3–4), 8–28.

Birch, J., Burn, C., Schnell, A., Browning, H., & Crump, A. (2021). *Review of the evidence of sentience in cephalopod molluscs and decapod crustaceans*. LSE Consulting.
Birch, J., Ginsburg, S., & Jablonka, E. (2020). Unlimited associative learning and the origins of consciousness: A primer and some predictions. *Biology & Philosophy, 35*(6), 1–23.
Birch, J., Schnell, A. K., & Clayton, N. S. (2020). Dimensions of animal consciousness. *Trends in Cognitive Sciences, 24*(10), 789–801.
Blackstone, N. W. (2001). Crustacea (crustaceans). *eLS*. https://doi.org/10.1002/9780470015902.a0001606.pub3
Block, N. (2004). Qualia. In R. L. Gregory (Ed.), *The Oxford companion to the mind* (pp. 785–789). Oxford University Press.
Boal, J. (2006). Social recognition: A top down view of cephalopod behaviour. *Vie et Milieu, 56*(2), 69–80.
Bonner, J. T. (1988). *The evolution of complexity*. Princeton University Press.
Boorse, C. (1977). Health as a theoretical concept. *Philosophy of Science, 44*(4), 542–573.
Braithwaite, V. (2010). *Do fish feel pain?* Oxford University Press.
Brandon, R. (1990). *Adaptation and environment*. Princeton University Press.
Bronfman, Z. Z., Ginsburg, S., & Jablonka, E. (2016). The transition to minimal consciousness through the evolution of associative learning. *Frontiers in Psychology, 7*, 1954.
Brooks, R. A. (1991). Intelligence without representation. *Artificial Intelligence, 47*, 139–159.
Broom, D. M. (2014). *Sentience and animal welfare*. CABI.
Browning, H. (2017). Anecdotes can be evidence too. *Animal Sentience, 16*(13).
Browning, H. (2019). What should we do about sheep? The role of intelligence in welfare considerations. *Animal Sentience, 25*(23).
Browning, H. (2020a). Assessing measures of animal welfare. https://link.springer.com/article/10.1007/s10539-022-09862-1
Browning, H. (2020b). *If I could talk to the animals: Measuring subjective animal welfare* [Ph.D. thesis]. Australian National University.
Browning, H. (2020c). The natural behavior debate: Two conceptions of animal welfare. *Journal of Applied Animal Welfare Science, 23*(3), 325–337.
Browning, H. (2022). The measurability of subjective animal welfare. *Journal of Consciousness Studies, 29*(3–4), 150–179.
Browning, H. (2023). Welfare comparisons within and across species. *Philosophical Studies, 180*, 529–551. https://doi.org/10.1007/s11098-022-01907-1
Browning, H., & Birch, J. (2022). Animal sentience. *Philosophy Compass, 17*(5), e12822. https://doi.org/10.1111/phc3.12822
Browning, H., & Veit, W. (2020a). Confined freedom and free confinement: The ethics of captivity in *Life of Pi*. In A. T. Bogár & R. S. Szigethy (Eds.), *Critical insights: Life of Pi* (pp. 119–134). Salem Press.
Browning, H., & Veit, W. (2020b). Improving invertebrate welfare. *Animal Sentience, 29*(4).
Browning, H., & Veit, W. (2020c). Is humane slaughter possible? *Animals, 10*(5), 799.
Browning, H., & Veit, W. (2020d). The measurement problem of consciousness. *Philosophical Topics, 48*(1), 85–108.
Browning, H., & Veit, W. (2021a). Evolutionary biology meets consciousness: Essay review of Simona Ginsburg and Eva Jablonka's *The Evolution of The Sensitive Soul*. *Biology & Philosophy, 36*(5). https://doi.org/10.1007/s10539-021-09781-7
Browning, H., & Veit, W. (2021b). Freedom and animal welfare. *Animals, 11*(4), 1148.
Browning, H., & Veit, W. (2021c). Positive wild animal welfare. *Preprint*. http://philsci-archive.pitt.edu/19608/
Browning, H., & Veit, W. (2022a). The importance of end-of-life welfare. *Animal Frontiers, 12*(1), 8–15.
Browning, H., & Veit, W. (2022b). On the relevance of experimental philosophy to neuroethics. *AJOB Neuroscience, 13*(1), 55–57.
Browning, H., & Veit, W. (2022c). The sentience shift in animal research. *The New Bioethics*, 1–16. https://doi.org/10.1080/20502877.2022.2077681
Browning, H., & Veit, W. (2022d). Autism and the preference for imaginary worlds. *Behavioral and Brain Sciences, 45*, E279.
Brunberg, E., Wallenbeck, A., & Keeling, L. J. (2011). Tail biting in fattening pigs: Associations between frequency of tail biting and other abnormal behaviours. *Applied Animal Behaviour Science, 133*(1–2), 18–25.

Bublitz, A., Weinhold, S. R., Strobel, S., Dehnhardt, G., & Hanke, F. D. (2017). Reconsideration of serial visual reversal learning in octopus (Octopus vulgaris) from a methodological perspective. *Frontiers in Physiology, 8*, 54.
Budd, G. E., & Telford, M. J. (2009). The origin and evolution of arthropods. *Nature, 457*(7231), 812–817.
Burgdorf, J., & Panksepp, J. (2006). The neurobiology of positive emotions. *Neuroscience & Biobehavioral Reviews, 30*(2), 173–187.
Burghardt, G. M. (1985). Animal awareness: Current perceptions and historical perspective. *American Psychologist, 40*(8), 905–919.
Cabanac, M. (1971). Physiological role of pleasure. *Science, 173*(4002), 1103–1107.
Cabanac, M. (1979). Sensory pleasure. *The Quarterly Review of Biology, 54*(1), 1–29.
Cabanac, M. (1992). Pleasure: The common currency. *Journal of Theoretical Biology, 155*(2), 173–200.
Cabanac, M. (1996). On the origin of consciousness, a postulate and its corollary. *Neuroscience & Biobehavioral Reviews, 20*(1), 33–40.
Cabanac, M. (1999). Emotion and phylogeny. *Journal of Consciousness Studies, 6*(67), 176–190.
Cabanac, M., Cabanac, A. J., & Parent, A. (2009). The emergence of consciousness in phylogeny. *Behavioural Brain Research, 198*(2), 267–272.
Cabanac, M., & Johnson, K. (1983). Analysis of a conflict between palatability and cold exposure in rats. *Physiology & Behavior, 31*(2), 249–253.
Cacioppo, J. T., & Gardner, W. L. (1999). Emotion. *Annual Review of Psychology, 50*(1), 191–214.
Call, J., & Tomasello, M. (2008). Does the chimpanzee have a theory of mind? 30 years later. *Trends in Cognitive Sciences, 12*(5), 187–192.
Calvo, P., Gagliano, M., Souza, G. M., & Trewavas, A. (2020). Plants are intelligent, here's how. *Annals of Botany, 125*(1), 11–28.
Campbell, N. (2001). What was Huxley's epiphenomenalism? *Biology & Philosophy, 16*(3), 357–375.
Canguilhem, G. (1991). *The normal and the pathological* (C. R. Fawcett, Trans.). Zone Books.
Caporael, L. R., Griesemer, J. R., & Wimsatt, W. C. (2014). *Developing scaffolds in evolution, culture, and cognition*. MIT Press.
Carel, H. (2007). Can I be ill and happy? *Philosophia, 35*(2), 95–110.
Carew, T. J., Walters, E. T., & Kandel, E. R. (1981). Associative learning in Aplysia: Cellular correlates supporting a conditioned fear hypothesis. *Science, 211*(4481), 501–504.
Carls-Diamante, S. (2017). The octopus and the unity of consciousness. *Biology & Philosophy, 32*(6), 1269–1287.
Carnap, R. (1950). *Logical foundations of probability*. University of Chicago Press.
Carruthers, P. (2018). Valence and value. *Philosophy and Phenomenological Research, 97*(3), 658–680.
Carruthers, P. (in press). On valence: Imperative or representation of value? *The British Journal for the Philosophy of Science.*
Caveney, S., W. Cladman, L. Verellen, & C. Donly (2006). Ancestry of neuronal monoamine transporters in the Metazoa. *Journal of Experimental Biology, 209*(24), 4858–4868.
Cazzolla Gatti, R., Velichevskaya, A., Gottesman, B., & Davis, K. (2021). Grey wolf may show signs of self-awareness with the sniff test of self-recognition. *Ethology Ecology & Evolution, 33*(4), 444–467.
Chalmers, D. J. (1995). Facing up to the problem of consciousness. *Journal of Consciousness Studies, 2*(3), 200–219.
Chalmers, D. J. (2020). Is the hard problem of consciousness universal? *Journal of Consciousness Studies, 27*(5–6), 227–257.
Chapman, R., & Veit, W. (2020). Representing the autism spectrum. *The American Journal of Bioethics, 20*(4), 46–48.
Chapman, R., & Veit, W. (2021). The essence of autism: Fact or artefact? *Molecular Psychology, 26*(5), 1440–1441.
Cheke, L. G., & Clayton, N. S. (2010). Mental time travel in animals. *Wiley Interdisciplinary Reviews: Cognitive Science, 1*(6), 915–930.
Cheke, L. G., & Clayton, N. S. (2012). Eurasian jays (Garrulus glandarius) overcome their current desires to anticipate two distinct future needs and plan for them appropriately. *Biology Letters, 8*(2), 171–175.
Christensen, M. S., Kristiansen, L., Rowe, J. B., & Nielsen, J. B. (2008). Action-blindsight in healthy subjects after transcranial magnetic stimulation. *Proceedings of the National Academy of Sciences, 105*(4), 1353–1357.

Churchland, P. S. (1988). Reduction and the neurobiological basis of consciousness. In A. J. Marcel & E. Bisiach (Eds.), *Consciousness in contemporary science* (pp. 273–304). Oxford University Press.
Churchland, P. S. (2002). *Brain-wise: Studies in neurophilosophy.* MIT Press.
Clark, A. (1997). *Being there: Putting brain, body, and world together again.* MIT Press.
Clark, R. E., Manns, J. R., & Squire, L. R. (2002). Classical conditioning, awareness, and brain systems. *Trends in Cognitive Sciences, 6*(12), 524–531.
Clayton, N. S., & Dickinson, A. (1998). Episodic-like memory during cache recovery by scrub jays. *Nature, 395*(6699), 272–274.
Clayton, N. S., Yu, K. S., & Dickinson, A. (2001). Scrub jays (Aphelocoma coerulescens) form integrated memories of the multiple features of caching episodes. *Journal of Experimental Psychology: Animal Behavior Processes, 27*(1), 17.
Clayton, N. S., Yu, K. S., & Dickinson, A. (2003). Interacting cache memories: Evidence of flexible memory use by scrub-jays. *Journal of Experimental Psychology: Animal Behavior Processes, 29*, 14–22.
Colwill, R. M., Absher, R. A., & Roberts, M. (1988). Context-US learning in Aplysia californica. *Journal of Neuroscience, 8*(12), 4434–4439.
Cowey, A. (2010). The blindsight saga. *Experimental Brain Research, 200*(1), 3–24.
Cox, J. J., Reimann, F., Nicholas, A. K., Thornton, G., Roberts, E., Springell, K., Karbani, G., Jafri, H., Mannan, J., Raashid, Y., Al-Gazali, L., Hamamy, H., Valente, E. M., Gorman, S., Williams, R., McHale, D. P., Wood, J. N., Gribble, F. M., & Woods, C. G. (2006). An SCN9A channelopathy causes congenital inability to experience pain. *Nature, 444*(7121), 894–898.
Crick, F., & Koch, C. (1990). Towards a neurobiological theory of consciousness. *Seminars in the Neurosciences, 2*, 263–275.
Crook, R. J. (2021). Behavioral and neurophysiological evidence suggests affective pain experience in octopus. *iScience, 24*(3), 102229.
Crook, R. J., & Walters, E. T. (2011). Nociceptive behavior and physiology of molluscs: Animal welfare implications. *ILAR Journal, 52*(2), 185–195.
Crump, A., G. Arnott, & E. J. Bethell (2018). Affect-driven attention biases as animal welfare indicators: Review and methods. *Animals, 8*(8), 136.
D'Eath, R. B. (1998). Can video images imitate real stimuli in animal behaviour experiments? *Biological Reviews, 73*(3), 267–292.
Dale, R., & Plotnik, J. M. (2017). Elephants know when their bodies are obstacles to success in a novel transfer task. *Scientific Reports, 7*, 46309.
Dally, J. M., Emery, N. J., & Clayton, N. S. (2005). Cache protection strategies by western scrub-jays, Aphelocoma californica: Implications for social cognition. *Animal Behaviour, 70*(6), 1251–1263.
Damasio, A. (1994). *Descartes' error: Emotion, reason, and the human brain.* Avon Books.
Damasio, A. R. (1999). *The feeling of what happens: Body, emotion and the making of consciousness.* Harcourt Brace.
Danckert, J., Striemer, C., & Rossetti, Y. (2021). Blindsight. In J. J. S. Barton & A. Leff (Eds.), *Handbook of clinical neurology* (Vol. 178, pp. 297–310). Elsevier.
Darwin, C. (1859). *On the origin of species by means of natural selection, or the preservation of favoured races in the struggle for life.* John Murray.
Darwin, C. (1871). *The descent of man and selection in relation to sex.* John Murray.
Darwin, C. (1872). *The expression of the emotions in man and animals.* John Murray.
Darwin, C., & Wallace, A. (1858). On the tendency of species to form varieties and on the perpetuation of varieties and species by natural means of selection. *Journal of the Proceedings of the Linnean Society of London. Zoology, 3*(9), 45–62.
Darwin, C. R. (1838). Notebook M: [Metaphysics on morals and speculations on expression]. *Darwin Online.* CUL-DAR125. Transcribed by Kees Rookmaaker and edited by Paul Barrett. http://darwin-online.org.uk/content/frameset?viewtype=side&itemID=CUL-DAR125.-&pageseq=1
Darwin, C. R. (1840). Old & useless notes about the moral sense & some metaphysical points. *Darwin Online.* CUL-DAR91.4-55. Transcribed and edited by Paul H. Barrett. http://darwin-online.org.uk/content/frameset?pageseq=57&itemID=CUL-DAR91.4-55&viewtype=text
Darwin, E. (1794). *Zoonomia; or, the laws of organic life* (Vol. 1). Johnson.
Dawkins, M. S. (2017a). Animal welfare with and without consciousness. *Talk at NYU 2017 Animal Consciousness Conference.* Retrieved March 12, 2021, https://www.youtube.com/playlist?list=PLY_s7b9LrR8UCcPqL59XuII68ALjgLt7

Dawkins, M. S. (2017b). Animal welfare with and without consciousness. *Journal of Zoology, 301*(1), 1–10.
Dawkins, M. S. (1990). From an animal's point of view: Motivation, fitness, and animal welfare. *Behavioural and Brain Sciences, 13*(1), 1–9.
Dawkins, M. S. (1998). Evolution and animal welfare. *The Quarterly Review of Biology, 73*(3), 305–328.
Dawkins, M. S. (2001). Who needs consciousness? *Animal Welfare, 10*, S19–S29.
Dawkins, M. S. (2021). *The science of animal welfare: Understanding what animals want*. Oxford University Press.
de Kort, S. R., Dickinson, A., & Clayton, N. S. (2005). Retrospective cognition by food-caching western scrub-jays. *Learning and Motivation, 36*(2), 159–176.
De Meester, G., & Baeckens, S. (2021). Reinstating reptiles: From clueless creatures to esteemed models of cognitive biology. *Behaviour, 158*(12–13), 1057–1076.
de Waal, F. (2016). *Are we smart enough to know how smart animals are?* WW Norton & Company.
de Waal, F. B. (2019). Fish, mirrors, and a gradualist perspective on self-awareness. *PLoS Biology, 17*(2), e3000112.
de Waal, F. B., & Ferrari, P. F. (2010). Towards a bottom-up perspective on animal and human cognition. *Trends in Cognitive Sciences, 14*(5), 201–207.
Deckel, A. W. (1995). Laterality of aggressive responses in *Anolis*. *Journal of Experimental Zoology, 272*(3), 194–200.
Deckel, A. W. (1997). Effects of alcohol consumption on lateralized aggression in *Anolis carolinensis*. *Brain Research, 756*(1–2), 96–105.
Dehaene, S. (2014). *Consciousness and the brain: Deciphering how the brain codes our thoughts*. Viking.
Dennett, D. C. (1973). Mechanism and responsibility. In T. Honderich (Ed.), *Essays on freedom of action* (pp. 159–184). Routledge.
Dennett, D. C. (1987). *The intentional stance*. MIT Press.
Dennett, D. C. (1991). *Consciousness explained*. Little, Brown and Co.
Dennett, D. C. (1995a). *Darwin's dangerous idea: Evolution and the meanings of life*. Simon and Schuster.
Dennett, D. C. (1995b). The selfish gene as a philosophical essay. In A. Grafen & M. Ridley (Eds.), *Richard Dawkins: How a scientist changed the way we think* (pp. 101–115). Oxford University Press.
Dennett, D. C. (1996). *Kinds of minds: Toward an understanding of consciousness*. MIT Press.
Dennett, D. C. (2005). *Sweet dreams: Philosophical obstacles to a science of consciousness*. MIT Press.
Dennett, D. C. (2017a). *From bacteria to Bach and Back: The evolution of minds*. W. W. Norton & Company.
Dennett, D. C. (2017b). Suffering matters. *Talk at NYU 2017 Animal Consciousness Conference*. Retrieved March 12, 2021, https://www.youtube.com/playlist?list=PLY_s7b9LrR8UCcPqL59XuIII68ALjgLt7
Dennett, D. C. (2019a). Review of other minds: The octopus, the sea and the deep origins of consciousness. *Biology & Philosophy, 34*(2), 1–6.
Dennett, D. C. (2019b). Welcome to strong illusionism. *Journal of Consciousness Studies, 26*(9–10), 48–58.
Denton, D. (2006). *The primordial emotions: The dawning of consciousness*. Oxford University Press.
Diamond, J. M. (1982). Big-bang reproduction and ageing in male marsupial mice. *Nature, 298*(5870), 115–116.
Dobson, F. S. (2013). Live fast, die young, and win the sperm competition. *Proceedings of the National Academy of Sciences, 110*(44), 17610–17611.
Dörrenberg, S., Rakoczy, H. & Liszkowski, U. (2018). How (not) to measure infant theory of mind: Testing the replicability and validity of four non-verbal measures. *Cognitive Development, 46*, 12–30.
Dretske, F. (1993). Conscious experience. *Mind, 102*(406), 263–283.
Dretske, F. (1999). The mind's awareness of itself. *Philosophical Studies: An International Journal for Philosophy in the Analytic Tradition, 95*(1/2), 103–124.
Dunlap, K. (1919). Are there any instincts? *Journal of Abnormal Psychology, 14*, 307–311.
Edelman, G. M., & Tononi, G. (2000). *A universe of consciousness: How matter becomes imagination*. Basic Books.
Eisemann, C., Jorgensen, W., Merritt, D., Rice, M., Cribb, B., Webb, P., & Zalucki, M. (1984). Do insects feel pain?—A biological view. *Experientia, 40*(2), 164–167.
Elwood, R. W. (2012). Evidence for pain in decapod crustaceans. *Animal Welfare, 21*(S2), 23–27.
Elwood, R. W., & Appel, M. (2009). Pain experience in hermit crabs? *Animal Behaviour, 77*(5), 1243–1246.
Emery, N. J., & Clayton, N. S. (2001). Effects of experience and social context on prospective caching strategies by scrub jays. *Nature, 414*(6862), 443–446.

Ereshefsky, M. (2009). Defining 'health' and 'disease'. *Studies in History and Philosophy of Science Part C: Studies in History and Philosophy of Biological and Biomedical Sciences, 40*(3), 221–227.
Escobar, J. M. (2012). Autopoiesis and Darwinism. *Synthese, 185*(1), 53–72.
Feinberg, T., & Mallatt, J. (2016). *The ancient origins of consciousness.* MIT Press.
Figdor, C. (2018). *Pieces of mind: The proper domain of psychological predicates.* Oxford University Press.
Figdor, C. (2020). Relationship between cognition and moral status needs overhaul. *Animal Sentience, 29*(3).
Finn, J. K., Tregenza, T. & Norman, M. D. (2009). Defensive tool use in a coconut-carrying octopus. *Current Biology, 19*(23), R1069–R1070.
Flanagan, O. J. (1991). *The science of the mind.* MIT Press.
Frankish, K. (2017). *Illusionism: As a theory of consciousness.* Imprint Academic.
Fritz, R. S., Stamp, N. E., & Halverson, T. G. (1982). Iteroparity and semelparity in insects. *The American Naturalist, 120*(2), 264–268.
Gagliano, M. (2017). The mind of plants: Thinking the unthinkable. *Communicative & Integrative Biology, 10*(2), 38427.
Gagliano, M. (2018). Inside the vegetal mind: On the cognitive abilities of plants. In F. Baluska, M. Gagliano, & G. Witzany (Eds.), *Memory and Learning in Plants* (pp. 215–220). Springer.
Gagliano, M., Vyazovskiy, V. V., Borbély, A. A., Grimonprez, M., & Depczynski, M. (2016). Learning by association in plants. *Scientific Reports, 6*(1), 1–9.
Gallagher, S. (2011). *The oxford handbook of the self.* Oxford University Press.
Gallup, G. G. (1970). Chimpanzees: Self-recognition. *Science, 167*(3914), 86–87.
Gallup, G. G., Jr., & Anderson, J. R. (2020). Self-recognition in animals: Where do we stand 50 years later? Lessons from cleaner wrasse and other species. *Psychology of Consciousness: Theory, Research, and Practice,* 7(1), 46.
Gibbons, M., & Sarlak, S. (2020). Inhibition of pain or response to injury in invertebrates and vertebrates. *Animal Sentience, 5*(29), 34.
Gibbons, M., Versace, E., Crump, A., Baran, B., & Chittka, L. (2022). Motivational trade-offs and modulation of nociception in bumblebees. *Proceedings of the National Academy of Sciences, 119*(31), e2205821119.
Ginsburg, S., & Jablonka, E. (2010). Experiencing: A Jamesian approach. *Journal of Consciousness Studies, 17*(5–6), 102–124.
Ginsburg, S., & Jablonka, E. (2019). *The evolution of the sensitive soul: Learning and the origins of consciousness.* MIT Press.
Godfrey-Smith, P. (1996a). *Complexity and the function of mind in nature.* Cambridge University Press.
Godfrey-Smith, P. (1996b). Précis of complexity and the function of mind in nature. *Adaptive Behavior, 4*(3–4), 453–465.
Godfrey-Smith, P. (1996c). Replies to four critics. *Adaptive Behavior, 4*(3–4), 486–493.
Godfrey-Smith, P. (1997). Author's response. *Metascience, 6*(2), 31–37.
Godfrey-Smith, P. (2002). Environmental complexity and the evolution of cognition. In R. Sternberg & J. Kaufman (Eds.), *The evolution of intelligence* (pp. 223–249). Lawrence Erlbaum.
Godfrey-Smith, P. (2013a). On the relation between philosophy and science. *Unpublished lecture manuscript from the Gesellschaft für Wissenschaftsphilosophie (GWP) Conference, Hannover 2013.* Retrieved June 23, 2020, https://www.petergodfreysmith.com/PhilosophyScience_PGS_2013.pdf
Godfrey-Smith, P. (2013b). *Philosophy of biology.* Princeton University Press.
Godfrey-Smith, P. (2016a). Animal evolution and the origins of experience. In D. Livingstone Smith (Ed.), *How biology shapes philosophy: New foundations for naturalism* (pp. 51–71). Cambridge University Press.
Godfrey-Smith, P. (2016b). Individuality, subjectivity, and minimal cognition. *Biology & Philosophy, 31*(6), 775–796.
Godfrey-Smith, P. (2016c). Mind, matter, and metabolism. *The Journal of Philosophy, 113*(10), 481–506.
Godfrey-Smith, P. (2016d). *Other minds: The octopus, the sea, and the deep origins of consciousness.* Farrar, Straus and Giroux.
Godfrey-Smith, P. (2017a). Animal evolution and subjective experience. *Talk at NYU 2017 Animal Consciousness Conference.* Retrieved March 12, 2021, https://www.youtube.com/playlist?list=PLY_s7b9LrR8UCcPqL59XuIII68ALjgLt7

Godfrey-Smith, P. (2017b). Complexity revisited. *Biology & Philosophy, 32*(3), 467–479.

Godfrey-Smith, P. (2017c). The subject as cause and effect of evolution. *Interface Focus,* 7(5), 20170022.

Godfrey-Smith, P. (2019a). Evolving across the explanatory gap. *Philosophy, Theory, and Practice in Biology, 11(1).* https://doi.org/10.3998/ptpbio.16039257.0011.001

Godfrey-Smith, P. (2019b). Octopus experience. *Animal Sentience, 26*(18).

Godfrey-Smith, P. (2020a). Gradualism and the evolution of experience. *Philosophical Topics, 48*(1), 201–220.

Godfrey-Smith, P. (2020b). *Metazoa: Animal life and the birth of the mind.* Farrar, Straus and Giroux.

Godfrey-Smith, P. (2020c). Varieties of subjectivity. *Philosophy of Science, 87*(5), 1150–1159.

Godfrey-Smith, P. (2021a). Integration, lateralization, and animal experience. *Mind & Language, 36*(2), 285–296.

Godfrey-Smith, P. (2021b). Learning and the biology of consciousness: A commentary on Birch, Ginsburg, and Jablonka. *Biology & Philosophy, 36*(5), 1–4.

Goff, P., W. Seager, & S. Allen-Hermanson (2020). Panpsychism. In E. N. Zalta (Ed.), *The Stanford encyclopedia of philosophy* (Summer 2020 ed.). Retrieved February 17, 2021, https://plato.stanford.edu/archives/sum2020/entries/panpsychism/

Goldman, A. I. (1993). Consciousness, folk psychology, and cognitive science. *Consciousness and Cognition, 2*(4), 364–382.

Gould, S. J. (1996). In the mind of the beholder. In S. J. Gould (Ed.), *Dinosaur in a haystack* (pp. 93–107). Jonathan Cape.

Gould, S. J., & Lewontin, R. C. (1979). The spandrels of San Marco and the Panglossian paradigm: A critique of the adaptationist programme. *Proceedings of The Royal Society of London. Series B. Biological Sciences, 205*(1161), 581–598.

Graziano, M. S. (2016). Consciousness engineered. *Journal of Consciousness Studies, 23*(11–12), 98–115.

Griesemer, J. (2014). Reproduction and scaffolded developmental processes: An integrated evolutionary perspective. In A. Minelli & T. Pradeu (Eds.), *Towards a theory of development* (pp. 183–202). University Press Oxford.

Griffin, D. R. (1976). *The question of animal awareness: Evolutionary continuity of mental experience.* Rockefeller University Press.

Griffin, D. R. (1981). *The question of animal awareness: Evolutionary continuity of mental experience* (revised and enlarged ed.). Rockefeller University Press.

Griffin, D. R. (1992). *Animal minds.* University of Chicago Press.

Griffin, D. R. (1998). From cognition to consciousness. *Animal Cognition, 1*(1), 3–16.

Griffin, D. R. (2001). *Animal minds: Beyond cognition to consciousness.* University of Chicago Press.

Griffiths, P. E. (1997). *What emotions really are: The problem of psychological categories.* University of Chicago Press.

Griffiths, P. E. (2008). Ethology, sociobiology, and evolutionary psychology. In S. Sarkar & A. Plutynski (Eds.), *A companion to the philosophy of biology* (pp. 393–414). Blackwell Publishing Ltd.

Griffiths, P. E. (2009). In what sense does 'nothing make sense except in the light of evolution'? *Acta Biotheoretica, 57*(1), 11–32.

Griffiths, P. E. (2017). Emotions. In W. Bechtel & G. Graham (Eds.), *A companion to cognitive science* (pp. 197–203). Blackwell Publishing Ltd.

Griffiths, P. E. (2018). What is an organism, and what is it for? Neo-Aristotelian, Darwinian and post-Hamilton perspectives. *Talk at PhilInBioMed Network Bordeaux.* Retrieved February 20, 2021, https://youtu.be/TPuYjYT_eKo

Griffiths, P. E., & Gray, R. D. (2001). Darwinism and developmental systems. In S. Oyama, P. Griffiths, & R. D. Gray (Eds.), *Cycles of contingency: Developmental systems and evolution* (pp. 195–218). MIT Press.

Griffiths, P. E., & Matthewson, J. (2018). Evolution, dysfunction, and disease: A reappraisal. *The British Journal for the Philosophy of Science, 69*(2), 301–327.

Groening, J., Venini, D., & Srinivasan, M. V. (2017). In search of evidence for the experience of pain in honeybees: A self-administration study. *Scientific Reports,* 7(1), 1–8.

Güntürkün, O., & Bugnyar, T. (2016). Cognition without cortex. *Trends in Cognitive Sciences, 20*(4), 291–303.

Haeckel, E. (1892). Our monism. The principles of a consistent, unitary world-view. *The Monist, 2*(4), 481–486.

Hanlon, R. (2007). Cephalopod dynamic camouflage. *Current Biology, 17*(11), R400–R404.
Hanlon, R. T., Chiao, C., Mäthger, L., Buresch, K. C., Barbosa, A., Allen, J. J., Siemann, L., & Chubb, C. (2011). Rapid adaptive camouflage in cephalopods. In M. Stevens & S. Merilaita (Eds.), *Animal Camouflage: Mechanisms and functions* (pp. 145–163). Cambridge University Press.
Hanlon, R. T., Watson, A. C., & Barbosa, A. (2010). A "mimic octopus" in the Atlantic: Flatfish mimicry and camouflage by Macrotritopus defilippi. *The Biological Bulletin, 218*(1), 15–24.
Hare, B., Call, J., Agnetta, B., & Tomasello, M. (2000). Chimpanzees know what conspecifics do and do not see. *Animal Behaviour, 59*(4), 771–785.
Healy, K., McNally, L., Ruxton, G. D., Cooper, N., & Jackson, A. L. (2013). Metabolic rate and body size are linked with perception of temporal information. *Animal Behaviour, 86*(4), 685–696.
Hentschel, E., & Penzlin, H. (1982). Beeinflussung des Putzverhaltens bei *Periplaneta americana* (L.) durch Wundsetzung, Naloxon-, Morphin-und MetEnkephalingaben. *Zoologische Jahrbücher. Abteilung fur allgemeine Zoologie und Physiologie der Tiere, 86*, 361–370.
Hesslow, G. (1993). Do we need a concept of disease? *Theoretical Medicine, 14*(1), 1–14.
Hill, C. S. (2018). Unity of consciousness. *Wiley Interdisciplinary Reviews: Cognitive Science, 9*(5), e1465.
Ho, M.-W., & Saunders, P. T. (1979). Beyond neo-Darwinism—An epigenetic approach to evolution. *Journal of Theoretical Biology, 78*(4), 573–591.
Houston, A. I., & McNamara, J. M. (1999). *Models of adaptive behaviour: An approach based on state*. Cambridge University Press.
Huebner, B. (2012). Reflection, reflex, and folk intuitions. *Consciousness and Cognition, 21*(2), 651–653.
Huffard, C. L. (2006). Locomotion by Abdopus aculeatus (Cephalopoda: Octopodidae): Walking the line between primary and secondary defenses. *Journal of Experimental Biology, 209*(19), 3697–3707.
Humphrey, N. (1992). *A history of the mind: Evolution and the birth of consciousness*. Simon and Schuster.
Humphrey, N. (2011). *Soul dust: The magic of consciousness*. Princeton University Press.
Hurley, S. L. (1998). *Consciousness in action*. Harvard University Press.
Huxley, T. H. (2011). On the hypothesis that animals are automata, and its history. In T. H. Huxley (Ed.), *Collected essays: Volume 1. Methods and results* (pp. 199–250). Cambridge University Press.
Huxley, T. H., & Youmans, W. J. (1868). *The elements of physiology and hygiene: A text-book for educational institutions*. Appleton & Co.
Ichikawa, J. J., & M. Steup (2018). The analysis of knowledge. In E. N. Zalta (Ed.), *The Stanford encyclopedia of philosophy* (Summer 2018 ed.). Retrieved January 18, 2020, https://plato.stanford.edu/archives/sum2018/entries/knowledge-analysis/
James, W. (1890). *The principles of psychology*. Dover Publications.
Jékely, G., Keijzer, F., & Godfrey-Smith, P. (2015). An option space for early neural evolution. *Philosophical Transactions of the Royal Society B: Biological Sciences, 370*(1684), 20150181.
Jennings, H. S. (1904). *Behavior of the lower organisms*. Columbia University Press.
Jolij, J., & Lamme, V. A. (2005). Repression of unconscious information by conscious processing: Evidence from affective blindsight induced by transcranial magnetic stimulation. *Proceedings of the National Academy of Sciences, 102*(30), 10747–10751.
Josef, N., Amodio, P., Fiorito, G., & Shashar, N. (2012). Camouflaging in a complex environment—Octopuses use specific features of their surroundings for background matching. *PLoS One*, 7(5), e37579.
Kabadayi, C., & Osvath, M. (2017). Ravens parallel great apes in flexible planning for tool-use and bartering. *Science, 357*(6347), 202–204.
Kano, F., Krupenye, C., Hirata, S., Tomonaga, M., & Call, J. (2019). Great apes use self-experience to anticipate an agent's action in a false-belief test. *Proceedings of the National Academy of Sciences, 116*(42), 20904–20909.
Kappeler, P. M. (2021). *Animal behaviour: An evolutionary perspective*. Springer.
Karg, K., Schmelz, M., Call, J., & Tomasello, M. (2015). The goggles experiment: Can chimpanzees use self-experience to infer what a competitor can see? *Animal Behaviour, 105*, 211–221.
Keefner, A. (2016). Corvids infer the mental states of conspecifics. *Biology & Philosophy, 31*(2), 267–281.
Keeley, B. L. (2002). Making sense of the senses: Individuating modalities in humans and other animals. *The Journal of Philosophy, 99*(1), 5–28.
Keenan, J. P., Gallup, G. G., & Falk, D. (2003). *The face in the mirror: The search for the origins of consciousness*. HarperCollins Publishers.

Keijzer, F. (2015). Moving and sensing without input and output: Early nervous systems and the origins of the animal sensorimotor organization. *Biology & Philosophy, 30*(3), 311–331.

Keijzer, F., & Arnellos, A. (2017). The animal sensorimotor organization: A challenge for the environmental complexity thesis. *Biology & Philosophy, 32*(3), 421–441.

Keijzer, F., Van Duijn, M., & Lyon, P. (2013). What nervous systems do: Early evolution, input–output, and the skin brain thesis. *Adaptive Behavior, 21*(2), 67–85.

Keogh, R., & J. Pearson (2018). The blind mind: No sensory visual imagery in aphantasia. *Cortex, 105*, 53–60.

Key, B. (2016). Why fish do not feel pain. *Animal Sentience, 1*(3), 1–34.

Kirk, R. (2021). Zombies. In E. N. Zalta (Ed.), *The Stanford encyclopedia of philosophy* (Spring 2021 ed.). Retrieved December 4, 2021, https://plato.stanford.edu/archives/spr2021/entries/zombies/

Klein, C., & Barron, A. B. (2020). How experimental neuroscientists can fix the hard problem of consciousness. *Neuroscience of Consciousness 2020*(1), niaa009.

Knobe, J., & Nichols, S. (2017). Experimental philosophy. In E. N. Zalta (Ed.), *The Stanford encyclopedia of philosophy* (Winter 2017 ed.). Retrieved February 28, 2021, https://plato.stanford.edu/archives/win2017/entries/experimental-philosophy/

Koch, C., & Tononi, G. (2011). A test for consciousness. *Scientific American 304*(6), 44–47.

Kohda, M., Hotta, T., Takeyama, T., Awata, S., Tanaka, H., Asai, J., & Jordan, A. L. (2019). If a fish can pass the mark test, what are the implications for consciousness and self-awareness testing in animals? *PLoS Biology, 17*(2), e3000021.

Kohda, M., Sogawa, S., Jordan, A. L., Kubo, N., Awata, S., Satoh, S., Kobayashi, T., Fujita, A., & Bshary, R. (2022). Further evidence for the capacity of mirror self-recognition in cleaner fish and the significance of ecologically relevant marks. *PLoS Biology, 20*(2), e3001529.

Krachun, C., Carpenter, M., Call, J., & Tomasello, M. (2009, July). A competitive nonverbal false belief task for children and apes. *Developmental Science, 12*(4), 521–535.

Kringelbach, M., & Berridge, K. (2010). *Pleasures of the brain*. Oxford University Press.

Krupenye, C., & Call, J. (2019). Theory of mind in animals: Current and future directions. *Wiley Interdisciplinary Reviews: Cognitive Science, 10*(6), e1503.

Kuo, Z. Y. (1921). Giving up instincts in psychology. *Journal of Philosophy, 18*, 645–664.

Kutschera, U., Levit, G. S., & Hossfeld, U. (2019). Ernst Haeckel (1834–1919): The German Darwin and his impact on modern biology. *Theory in Biosciences, 138*(1), 1–7.

Labandeira, C. C. (2006). Silurian to Triassic plant and insect clades and their associations: New data, a review, and interpretations. *Arthropod Systematics & Phylogeny, 64*, 53–94.

Lack, D. (1947). The significance of clutch-size. *Ibis, 89*(2), 302–352.

Laland, K., Uller, T., Feldman, M., Sterelny, K., Müller, G. B., Moczek, A., Jablonka, E., Odling-Smee, J., Wray, G. A., Hoekstra, H. E., Futuyma, D. J., Lenski, R. E., Mackay, T. F. C., Schluter, D., & Strassmann, J. E.. (2014). Does evolutionary theory need a rethink? *Nature News, 514*(7521), 161.

Lamarck, J. (1984). *Zoological philosophy* (H. Elliot, Trans.). University of Chicago Press.

Lambert, H., Carder, G., & D'Cruze, N. (2019). Given the cold shoulder: A review of the scientific literature for evidence of reptile sentience. *Animals, 9*(10), 821.

Lambert, H., Elwin, A., & D'Cruze, N. (2021). Wouldn't hurt a fly? A review of insect cognition and sentience in relation to their use as food and feed. *Applied Animal Behaviour Science, 243*, 105432.

Laughlin, S. B., & Weckström, M. (1993). Fast and slow photoreceptors—A comparative study of the functional diversity of coding and conductances in the Diptera. *Journal of Comparative Physiology A, 172*(5), 593–609.

Lea, S. E., & Dittrich, W. H. (2000). What do birds see in moving video images? In J. Fagot (Ed.), Picture *p*erception in *a*nimals (pp. 143–180). Psychology Press. https://www.amazon.co.uk/Picture-Perception-Animals-Joel-Fagot/dp/1841691569

Learmonth, M. J. (2020). The matter of non-avian reptile sentience, and why it "matters" to them: A conceptual, ethical and scientific review. *Animals, 10*(5), 901.

LeDoux, J. E. (2019). *The deep history of ourselves: The four-billion-year story of how we got conscious brains*. Viking.

LeDoux, J. E. (2017a). Emotional consciousness in animals: The case of fearful feelings. *Talk at NYU 2017 Animal Consciousness Conference*. Retrieved March 12, 2021, https://www.youtube.com/playlist?list=PLY_s7b9LrR8UCcPqL59XuIII68ALjgLt7

LeDoux, J. E. (2017b). Semantics, surplus meaning, and the science of fear. *Trends in Cognitive Sciences, 21*(5), 303–306.

LeDoux, J. E. (2022). As soon as there was life, there was danger: The deep history of survival behaviours and the shallower history of consciousness. *Philosophical Transactions of the Royal Society B, 377*(1844), 20210292.

Lee, A. Y. (in press). Degrees of Consciousness. *Noûs*. https://onlinelibrary.wiley.com/doi/full/10.1111/nous.12421

Levine, J. (1983). Materialism and qualia: The explanatory gap. *Pacific Philosophical Quarterly, 64*(4), 354–361.

Levins, R., & Lewontin, R. C. (1985). *The dialectical biologist*. Harvard University Press.

Levy, D. J., & Glimcher, P. (2016). Common value representation—A neuroeconomics perspective. In T. Brosch & D. Sander (Eds.), *Handbook of value: Perspectives from economics, neuroscience, philosophy, psychology and sociology* (pp. 85–118). Oxford University Press.

Levy, D. J., & Glimcher, P. W. (2012). The root of all value: A neural common currency for choice. *Current Opinion in Neurobiology, 22*(6), 1027–1038.

Levy, G., & Hochner, B. (2017). Embodied organization of Octopus vulgaris morphology, vision, and locomotion. *Frontiers in Physiology, 8*, 164.

Lewontin, R. C. (1979). Sociobiology as an adaptationist program. *Behavioral Science, 24*(1), 5–14.

Lewontin, R. C. (1981). On constraints and adaptation. *Behavioral and Brain Sciences, 4*(2), 244–245.

Lewontin, R. C. (1985). The organism as the subject and object of evolution. In R. Levins & R. C. Lewontin (Eds.), *The dialectical biologist*. Harvard University Press.

Lewontin, R. C. (1998). The evolution of cognition: Questions we will never answer. In D. Scarborough & S. Sternberg (Eds.), *An invitation to cognitive science* (Vol. 4, pp. 107–132). MIT Press.

Lewontin, R. C., & Levins, R. (1997). Organism and environment. *CNS, 8*(2), 95–98.

Liljenström, H., & Århem, P. (2008). *Consciousness transitions: Phylogenetic, ontogenetic and physiological aspects*. Elsevier.

Lorenz, K. (1950). The comparative method in studying innate behavior patterns. In *Physiological mechanisms in animal behavior (Society's Symposium IV)* (pp. 221–268). Academic Press.

Lorenz, K. (1981). *The foundations of ethology*. Springer.

Lorenz, K. (1997). *The natural science of the human species: An introduction to comparative behavioral research: The "Russian Manuscript" (1944–1948)*. Translated by R. D. Martin and edited by A. von Cranach. MIT Press.

Lyon, P. (2006). *The agent in the organism: Towards a biogenic theory of cognition* [Ph.D. thesis]. The Australian National University.

Machery, E., & Sytsma, J. (2011). Robot pains and corporate feelings. *The Philosophers' Magazine, 52*(52), 78–82.

Macphail, E. (1998). *The evolution of consciousness*. Oxford University Press.

Maloof, A. C., Porter, S. M., Moore, J. L., Dudás, F. O., Bowring, S. A., Higgins, J. A., Fike, D. A., & Eddy, M. P. (2010). The earliest Cambrian record of animals and ocean geochemical change. *GSA Bulletin, 122*(11–12), 1731–1774. https://pubs.geoscienceworld.org/gsa/gsabulletin/article-abstract/122/11-12/1731/125474/The-earliest-Cambrian-record-of-animals-and-ocean?redirectedFrom=fulltext

Mangel, M., & Clark, C. W. (1986). Towards a unifield foraging theory. *Ecology, 67*(5), 1127–1138.

Manus, M. B. (2018). Evolutionary mismatch. *Evolution, Medicine, and Public Health, 2018*(1), 190–191.

Margulis, L. (2001). The conscious cell. *Annals of the New York Academy of Sciences, 929*(1), 55–70.

Marshall, R., & Godfrey-Smith, P. (2014). Philosophy of biology: Peter Godfrey-Smith interviewed by Richard Marshall. *3:16 Blog*. Retrieved January 2, 2022, https://www.3-16am.co.uk/articles/philosophy-of-biology

Marshall, N. J., & Messenger, J. B. (1996). Colour-blind camouflage. *Nature, 382*(6590), 408–409.

Maruzzo, D., & Bortolin, F. (2013). Arthropod regeneration. In A. Minelli, G. Boxshall, & G. Fusco (Eds.), *Arthropod biology and evolution* (pp. 149–169). Springer.

Mascetti, G. G. (2016). Unihemispheric sleep and asymmetrical sleep: Behavioral, neurophysiological, and functional perspectives. *Nature and Science of Sleep, 8*, 221.

Mashour, G. A., & Alkire, M. T. (2013). Evolution of consciousness: Phylogeny, ontogeny, and emergence from general anesthesia. *Proceedings of the National Academy of Sciences, 110*(Supplement 2), 10357–10364.

Mather, J. (2019). What is in an octopus's mind? *Animal Sentience, 26*(1).
Mather, J. (2021a). The case for octopus consciousness: Unity. *NeuroSci, 2*(4), 405–415.
Mather, J. (2021b). Octopus consciousness: The role of perceptual richness. *NeuroSci, 2*(3), 276–290.
Matthewson, J. (2020). Does proper function come in degrees? *Biology & Philosophy, 35*(4), 1–18.
Matthewson, J., & Griffiths, P. E. (2017). Biological criteria of disease: Four ways of going wrong. *Journal of Medicine and Philosophy, 42*(4), 447–466.
Maturana, H. R., & Varela, F. J. (1980). *Autopoiesis and cognition: The realization of the living*. D. Reidel Publishing Company.
Maynard Smith, J. (1952). The importance of the nervous system in the evolution of animal flight. *Evolution, 6*(1), 127–129.
Maynard Smith, J. (1987). Evolutionary progress and levels of selection. In J. Dupré (Ed.), *The latest on the best: Essays on evolution and optimality* (pp. 219–230). MIT Press.
Mayr, E. (1988). *Toward a new philosophy of biology: Observations of an evolutionist*. Harvard University Press.
McCleery, R. (1977). On satiation curves. *Animal Behaviour, 25*, 1005–1015.
McFarland, D., & Sibly, R. (1975). The behavioural final common path. *Philosophical Transactions of the Royal Society of London. B, Biological Sciences, 270*(907), 265–293.
McNamara, J. M. (1997). Optimal life histories for structured populations in fluctuating environments. *Theoretical Population Biology, 51*(2), 94–108.
McNamara, J. M., & Houston, A. I. (1986). The common currency for behavioral decisions. *The American Naturalist, 127*(3), 358–378.
McNamara, J. M., & Houston, A. I. (1996). State-dependent life histories. *Nature, 380*(6571), 215–221.
McNamara, J. M., & Houston, A. I. (2008). Optimal annual routines: Behaviour in the context of physiology and ecology. *Philosophical Transactions of the Royal Society B: Biological Sciences, 363*(1490), 301–319.
McNamara, J. M., & Houston, A. I. (2009). Integrating function and mechanism. *Trends in Ecology & Evolution, 24*(12), 670–675.
Melis, A. P., Call, J. & Tomasello, M. (2006). Chimpanzees (Pan troglodytes) conceal visual and auditory information from others. *Journal of Comparative Psychology, 120*(2), 154.
Meltzoff, A. N. (2007). 'Like me': A foundation for social cognition. *Developmental Science, 10*(1), 126–134.
Mendl, M., Burman, O. H. P., & Paul, E. S. (2010). An integrative and functional framework for the study of animal emotion and mood. *Proceedings of the Royal Society B: Biological Sciences, 277*(1696), 2895–2904.
Merker, B. (2005). The liabilities of mobility: A selection pressure for the transition to consciousness in animal evolution. *Consciousness and Cognition, 14*(1), 89–114.
Merker, B. (2007). Consciousness without a cerebral cortex: A challenge for neuroscience and medicine. *Behavioral and Brain Sciences, 30*(1), 63–81.
Merker, B., Williford, K., & Rudrauf, D. (2021). The integrated information theory of consciousness: A case of mistaken identity. *Behavioral and Brain Sciences, 45*, E41.
Merker, B., Williford, K., & Rudrauf, D. (2022). The integrated information theory of consciousness: Unmasked and identified. *Behavioral and Brain Sciences, 45*, E65.
Messenger, J. B. (1977). Evidence that octopus is colour blind. *Journal of Experimental Biology, 70*(1), 49–55.
Metz, J. A., Nisbet, R. M., & Geritz, S. A. (1992). How should we define 'fitness' for general ecological scenarios? *Trends in Ecology & Evolution, 7*(6), 198–202.
Meyer, A. (1895). How was Wallace led to the discovery of natural selection? *Nature, 52*(1348), 415–415.
Mikhalevich, I., & Powell, R. (2020). Minds without spines: Evolutionarily inclusive animal ethics. *Animal Sentience, 29*(1).
Millikan, R. G. (1995). *White queen psychology and other essays for Alice*. MIT Press.
Millikan, R. G. (2002). Biofunctions: Two paradigms. In R. Cummins, A. Ariew, &, M. Perlman (Eds.), *Functions: New readings in the philosophy of psychology and biology* (pp. 113–143). Oxford University Press.
Misof, B., Liu, S., Meusemann, K., Peters, R. S., Donath, A., Mayer, C., Frandsen, P. B., Ware, J. Flouri, T., Beutel, R. G., Niehuis, O., Petersen, M., Izquierdo-Carrasco, F., Wappler, T., Rust, J., Aberer, A. J., Aspöck, U., Aspöck, H., Bartel, D., ... Zhou, X. (2014). Phylogenomics resolves the timing and pattern of insect evolution. *Science, 346*(6210), 763–767. https://www.science.org/doi/10.1126/science.1257570
Mitoh, S., & Yusa, Y. (2021). Extreme autotomy and whole-body regeneration in photosynthetic sea slugs. *Current Biology, 31*(5), R233–R234.

Morbeck, M., Galloway, A., & A. Zihlman (1997). *The evolving female: A life-history perspective*. Princeton University Press.

Morrison, R., & Reiss, D. (2018). Precocious development of self-awareness in dolphins. *PLoS One, 13*(1), e0189813.

Müller, G. B. (2017). Why an extended evolutionary synthesis is necessary. *Interface Focus,* 7(5), 20170015.

Murphy, D. (2020). Concepts of disease and health. In E. N. Zalta (Ed.), *The Stanford encyclopedia of philosophy* (Summer 2020 ed.). Retrieved July 20, 2020, https://plato.stanford.edu/archives/sum2020/entries/health-disease/

Nagasako, E. M., Oaklander, A. L., & Dworkin, R. H. (2003). Congenital insensitivity to pain: An update. *Pain, 101*(3), 213–219.

Nagel, T. (1974). What is it like to be a bat? *The Philosophical Review, 83*(4), 435–450.

Nagel, T. (1986). *The view from nowhere*. Oxford University Press.

Nagel, T. (2012). *Mind and cosmos: Why the materialist neo-Darwinian conception of nature is almost certainly false*. Oxford University Press.

Narins, P. M., Grabul, D. S., Soma, K. K., Gaucher, P., & Hödl, W. (2005). Crossmodal integration in a dart-poison frog. *Proceedings of the National Academy of Sciences, 102*(7), 2425–2429.

Nesher, N., Levy, G., Grasso, F. W., & Hochner, B. (2014). Self-recognition mechanism between skin and suckers prevents octopus arms from interfering with each other. *Current Biology, 24*(11), 1271–1275.

Nesse, R. M. (2007). The importance of evolution for medicine. In W. R. Trevathan, J. J. McKenna, & E. O. Smith (Eds.), *Evolutionary medicine* (2nd ed., pp. 416–432). Oxford University Press.

Nesse, R. M., & Williams, G. C. (1995). *Evolution and healing: The new science of Darwinian medicine*. Weidenfeld & Nicolson.

Neto, C. (2020). When imprecision is a good thing, or how imprecise concepts facilitate integration in biology. *Biology & Philosophy, 35*(6), 1–21.

Nettle, D., & Frankenhuis, W. E. (2020). Life-history theory in psychology and evolutionary biology: One research programme or two? *Philosophical Transactions of the Royal Society B, 375*(1803), 20190490.

Nichols, S., & Stich, S. P. (2003). *Mindreading: An integrated account of pretence, self-awareness, and understanding other minds*. Oxford University Press.

Nicholson, D. J. (2014). The return of the organism as a fundamental explanatory concept in biology. *Philosophy Compass, 9*(5), 347–359.

Nobel Prize Outreach (2021). Press release: The Nobel Prize in Physiology or Medicine 1973. *NobelPrize.org*. Retrieved June 30, 2021, https://www.nobelprize.org/prizes/medicine/1973/press-release/

Noble, D. (2015). Evolution beyond neo-Darwinism: A new conceptual framework. *Journal of Experimental Biology, 218*(1), 7–13.

Nordenfelt, L. (1995). *On the nature of health: An action-theoretic approach*. Kluwer.

O Connell, S., & Dunbar, R. (2003). A test for comprehension of false belief in chimpanzees. *Evolution and Cognition, 9*(2), 131–140.

Okasha, S. (2018). *Agents and goals in evolution*. Oxford University Press.

Ortega, L. J., Stoppa, K., Güntürkün, O., & Troje, N. F. (2008). Limits of intraocular and interocular transfer in pigeons. *Behavioural Brain Research, 193*(1), 69–78.

Ostojíc, L., Legg, E. W., Brecht, K. F., Lange, F., Deininger, C., Mendl, M., & S. Clayton, N. (2017). Current desires of conspecific observers affect cache-protection strategies in California scrub-jays and Eurasian jays. *Current Biology, 27*(2), R51–R53.

Osvath, M., & Sima, M. (2014). Sub-adult ravens synchronize their play: A case of emotional contagion. *Animal Behavior and Cognition, 1*(2), 197–205.

Paley, W. (1802). *Natural theology: Or, Evidences of the existence and attributes of the deity*. J. Faulder.

Palombo, D. J., Alain, C., Söderlund, H., Khuu, W., & Levine, B. (2015). Severely deficient autobiographical memory (SDAM) in healthy adults: A new mnemonic syndrome. *Neuropsychologia, 72*, 105–118.

Panksepp, J. (1998). *Affective neuroscience: The foundations of human and animal emotions*. Oxford University Press.

Panksepp, J. (2005). Affective consciousness: Core emotional feelings in animals and humans. *Consciousness and Cognition, 14*(1), 30–80.

Panksepp, J. (2010). Affective consciousness in animals: Perspectives on dimensional and primary process emotion approaches. *Proceedings of the Royal Society B: Biological Sciences, 277*(1696), 2905–2907.

Panksepp, J. (2011). Cross-species affective neuroscience decoding of the primal affective experiences of humans and related animals. *PLoS One, 6*(9), e21236.

Parkhaev, P. Y. (2007). The Cambrian 'basement' of gastropod evolution. *Geological Society, London, Special Publications, 286*(1), 415–421.
Pearce, J. M., Esber, G. R., George, D. N., & Haselgrove, M. (2008). The nature of discrimination learning in pigeons. *Learning & Behavior, 36*(3), 188–199.
Pearson, J. (2019). The human imagination: The cognitive neuroscience of visual mental imagery. *Nature Reviews Neuroscience, 20*(10), 624–634.
Pearson, J. M., Watson, K. K., & Platt, M. L. (2014). Decision making: The neuroethological turn. *Neuron, 82*(5), 950–965.
Pepperberg, I. M., Garcia, S. E., Jackson, E. C., & Marconi, S. (1995). Mirror use by African grey parrots (Psittacus erithacus). *Journal of Comparative Psychology, 109*(2), 182.
Peressini, A. (2014). Blurring two conceptions of subjective experience: Folk versus philosophical phenomenality. *Philosophical Psychology, 27*(6), 862–889.
Perry, C. J., Baciadonna, L., & Chittka, L. (2016). Unexpected rewards induce dopamine-dependent positive emotion-like state changes in bumblebees. *Science, 353*(6307), 1529–1531.
Perry, C. J., Barron, A. B. & Cheng, K. (2013). Invertebrate learning and cognition: Relating phenomena to neural substrate. *Wiley Interdisciplinary Reviews: Cognitive Science, 4*(5), 561–582.
Peterson, K. J., Cotton, J. A., Gehling, J. G., & Pisani, D. (2008). The Ediacaran emergence of bilaterians: Congruence between the genetic and the geological fossil records. *Philosophical Transactions of the Royal Society B: Biological Sciences, 363*(1496), 1435–1443.
Phillips, J., Buckwalter, W., Cushman, F., Friedman, O., Martin, A., Turri, J., Santos, L., & Knobe, J. (2020). Knowledge before belief. *Behavioral and Brain Sciences, 44*, E140.
Pianka, E. R., & Parker, W. S. (1975). Age-specific reproductive tactics. *The American Naturalist, 109*(968), 453–464.
Pirotta, E., Mangel, M., Costa, D. P., Mate, B., Goldbogen, J. A., Palacios, D. M., Hückstädt, L. A., McHuron, E. A., Schwarz, L., & New, L. (2018). A dynamic state model of migratory behavior and physiology to assess the consequences of environmental variation and anthropogenic disturbance on marine vertebrates. *The American Naturalist, 191*(2), E40–E56.
Pittendrigh, C. S. (1958). Adaptation, natural selection, and behavior. In A. Roe & G. G. Simpson (Eds.), *Behavior and evolution*. Yale University Press.
Plotnik, J. M., De Waal, F. B., & Reiss, D. (2006). Self-recognition in an Asian elephant. *Proceedings of the National Academy of Sciences, 103*(45), 17053–17057.
Plutchik, R. (1962). *The emotions: Facts, theories and a new model*. Random House.
Plutchik, R. (1980). A general psychoevolutionary theory of emotion. In R. Plutchik & H. Kellerman (Eds.), *Emotion: Theory, research, and experience* (pp. 3–33). Academic Press.
Potier, S., Lieuvin, M., Pfaff, M., & Kelber, A. (2020). How fast can raptors see? *Journal of Experimental Biology, 223*(1), jeb209031.
Premack, D., & Woodruff, G. (1978). Does the chimpanzee have a theory of mind? *Behavioral and Brain Sciences, 1*(4), 515–526.
Prinz, J. J. (2000). A neurofunctional theory of visual consciousness. *Consciousness and Cognition, 9*(2), 243–259.
Prinz, J. J. (2012). *The conscious brain: How attention engenders experience*. Oxford University Press.
Prior, H., Schwarz, A., & Güntürkün, O. (2008). Mirror-induced behavior in the magpie (Pica pica): Evidence of self-recognition. *PLoS Biology, 6*(8), e202.
Railo, H., & Hurme, M. (2021). Is the primary visual cortex necessary for blindsight-like behavior? Review of transcranial magnetic stimulation studies in neurologically healthy individuals. *Neuroscience & Biobehavioral Reviews, 127*, 353–364.
Rattenborg, N. C., Voirin, B., Cruz, S. M., Tisdale, R., Dell'Omo, G., Lipp, H.-P., Wikelski, M., & Vyssotski, A. L. (2016). Evidence that birds sleep in mid-flight. *Nature Communications, 7*(1), 1–9.
Reber, A. S. (2019). *The first minds: Caterpillars, karyotes, and consciousness*. Oxford University Press.
Rich, A. N., Bradshaw, J. L., & Mattingley, J. B. (2005). A systematic, large-scale study of synaesthesia: Implications for the role of early experience in lexical-colour associations. *Cognition, 98*(1), 53–84.
Ristau, C. A. (1992). Cognitive ethology: Past, present and speculations on the future. In *PSA: Proceedings of the Biennial Meeting of the Philosophy of Science Association* (Vol. 1992, pp. 125–136). Philosophy of Science Association.

Ro, T., Shelton, D., Lee, O. L., & Chang, E. (2004). Extrageniculate mediation of unconscious vision in transcranial magnetic stimulation-induced blindsight. *Proceedings of the National Academy of Sciences, 101*(26), 9933–9935.

Rodenburg, T., Buitenhuis, A., Ask, B., Uitdehaag, K., Koene, P., Van der Poel, J., & Bovenhuis, H. (2003). Heritability of feather pecking and open-field response of laying hens at two different ages. *Poultry Science, 82*(6), 861–867.

Roff, D. A. (1992). *Evolution of life histories: Theory and analysis.* Chapman and Hall.

Roff, D. A. (2002). *Life history evolution.* W. H. Freeman.

Rogers, L. J., Vallortigara, G., & Andrew, R. J. (2013). *Divided brains: The biology and behaviour of brain asymmetries.* Cambridge University Press.

Rolls, E. T. (1999). *The brain and emotion.* Oxford University Press.

Romanes, G. J. (1883). *Mental evolution in animals.* Kegan Paul, Trench & Co.

Russell, J. A. (2003). Core affect and the psychological construction of emotion. *Psychological Review, 110*(1), 145.

Russell, J. A., & Barrett, L. F. (1999). Core affect, prototypical emotional episodes, and other things called emotion: Dissecting the elephant. *Journal of Personality and Social Psychology, 76*(5), 805.

Russell, J. A., & Fehr, B. (1987). Relativity in the perception of emotion in facial expressions. *Journal of Experimental Psychology: General, 116*(3), 223.

Samuelson, L., & Swinkels, J. M. (2006). Information, evolution and utility. *Theoretical Economics, 1*(1), 119–142.

Saxe, R., & Kanwisher, N. (2003). People thinking about thinking people: The role of the temporo-parietal junction in "theory of mind". *Neuroimage, 19*(4), 1835–1842.

Schechter, E. (2018). *Self-consciousness and "Split" brains: The minds' I.* Oxford University Press.

Schnell, A. K., & Clayton, N. S. (2021). Cephalopods: Ambassadors for rethinking cognition. *Biochemical and Biophysical Research Communications, 564*, 27–36.

Schnell, A. K., Hanlon, R. T., Benkada, A., & Jozet-Alves, C. (2016). Lateralization of eye use in cuttlefish: Opposite direction for anti-predatory and predatory behaviors. *Frontiers in Physiology, 7*, 620.

Schukraft, J. (2020). Does critical flicker-fusion frequency track the subjective experience of time? *Rethink Priorities.* Retrieved May 6, 2022, https://rethinkpriorities.org/publications/does-critical-flicker-fusion-frequencytrack-the-subjective-experience-of-time

Schwartz, S. K., Wagner, W. E., Jr., & Hebets, E. A. (2016). Males can benefit from sexual cannibalism facilitated by self-sacrifice. *Current Biology, 26*(20), 2794–2799.

Sebo, J. (2018). The moral problem of other minds. *The Harvard Review of Philosophy, 25*, 51–70.

Sheets-Johnstone, M. (1999). *The primacy of movement.* John Benjamins Pub.

Shen, B., Dong, L., Xiao, S., & Kowalewski, M. (2008). The Avalon explosion: Evolution of Ediacara morphospace. *Science, 319*(5859), 81–84.

Shermer, M. (2002). *In Darwin's shadow: The life and science of Alfred Russel Wallace: A biographical study on the psychology of history.* Oxford University Press.

Sherrington, C. S. (1906). *The Integrative Action of the Nervous System.* Yale University Press.

Shizgal, P., & Conover, K. (1996). On the neural computation of utility. *Current Directions in Psychological Science, 5*(2), 37–43.

Sneddon, L. U., Lopez-Luna, J., Wolfenden, D. C., Leach, M. C., Valentim, A. M., Steenbergen, P. J., Bardine, N., Currie, A. D., Broom, D. M., & Brown, C. (2018). Fish sentience denial: Muddying the waters. *Animal Sentience, 3*(21), 1.

Solms, M. (2021). *The hidden spring: A journey to the source of consciousness.* W. W. Norton & Company.

Sovrano, V. A., Bisazza, A., & Vallortigara, G. (2001). Lateralization of response to social stimuli in fishes: A comparison between different methods and species. *Physiology & Behavior, 74*(1–2), 237–244.

Spencer, H. (1855). *Principles of psychology.* Longman, Brown and Green.

Spencer, H. (1870). *First principles.* Williams and Norgate.

Sprecher, S. G., Cardona, A., & Hartenstein, V. (2011). The Drosophila larval visual system: High-resolution analysis of a simple visual neuropil. *Developmental Biology, 358*(1), 33–43.

Spurrett, D. (2014). Philosophers should be interested in 'common currency' claims in the cognitive and behavioural sciences. *South African Journal of Philosophy, 33*(2), 211–221.

Spurrett, D. (2015). The natural history of desire. *South African Journal of Philosophy, 34*(3), 304–313.

Spurrett, D. (2020). The descent of preferences. *The British Journal for the Philosophy of Science* [Online First]. https://doi.org/10.1093/bjps/axz020

Stearns, S. C. (1992). *The evolution of life histories*. Oxford University Press.

Sterelny, K. (1997). Where does thinking come from? A commentary on Peter Godfrey-Smith's complexity and the function of mind in nature. *Biology & Philosophy, 12*(4), 551–566.

Sterelny, K. (2003). *Thought in a hostile world*. Blackwell.

Sterelny, K., & Griffiths, P. E. (1999). *Sex and death: An introduction to philosophy of biology*. University of Chicago Press.

Stoltzfus, A. (2019). Understanding bias in the introduction of variation as an evolutionary cause. In T. Uller & K. N. Laland (Eds.), *Evolutionary causation: Biological and philosophical reflections* (pp. 29–61). MIT Press.

Suzuki, Y., Chou, J., Garvey, S. L., Wang, V. R., & Yanes, K. O. (2019). Evolution and regulation of limb regeneration in arthropods. In W. Tworzydlo & S. M. Bilinski (Eds.), *Evo-devo: Non-model species in cell and developmental biology* (pp. 419–454). Springer.

Sytsma, J. (2010). Dennett's theory of the folk theory of consciousness. *Journal of Consciousness Studies, 17*(3–4), 107–130.

Sytsma, J. (2012). Revisiting the valence account. *Philosophical Topics, 40*(2), 179–198.

Sytsma, J., & Machery, E. (2009). How to study folk intuitions about phenomenal consciousness. *Philosophical Psychology, 22*(1), 21–35.

Sytsma, J., & Machery, E. (2010). Two conceptions of subjective experience. *Philosophical Studies, 151*(2), 299–327.

Sytsma, J., & Machery, E. (2012). On the relevance of folk intuitions: A commentary on Talbot. *Consciousness and Cognition, 21*(2), 654–660.

Sytsma, J., & E. Ozdemir (2019). No problem: Evidence that the concept of phenomenal consciousness is not widespread. *Journal of Consciousness Studies, 26*(9–10), 241–256.

Talbot, B. (2012). The irrelevance of dispositions and difficulty to intuitions about the "hard problem" of consciousness: A response to Sytsma, Machery, and Huebner. *Consciousness and Cognition, 21*(2), 661–666.

Tanaka, M. M., Godfrey-Smith, P., & Kerr, B. (2020). The dual landscape model of adaptation and niche construction. *Philosophy of Science, 87*(3), 478–498.

Thompson, E. (2007). *Mind in Life: Biology, Phenomenology, and the Sciences of Mind*. Harvard University Press.

Thompson, E. (2011). Reply to commentaries. *Journal of Consciousness Studies, 18*(5–6), 176–223.

Thompson, E. (2022). Could all life be sentient? *Journal of Consciousness Studies, 29*(3–4), 229–265.

Thompson, N. S. (1986a). Deception and the concept of behavioral design. In R. W. Mitchell & N. Thompson (Eds.), *Deception* (pp. 53–66). SUNY Press.

Thompson, N. S. (1986b). Ethology and the birth of comparative teleonomy. In R. Campan & R. Zayan (Eds.), *Relevance of models and theory in ethology* (pp. 13–23). Privat IEC.

Thrush, D. (1973). Congenital insensitivity to pain. *Brain, 96*(2), 369–386.

Todd, J. T., & E. K. Morris (1986). The early research of John B. Watson: Before the behavioral revolution. *The Behavior Analyst, 9*(1), 71–88.

Tolman, E. (1923). The nature of instinct. *Psychological Bulletin, 20*(4), 200–218.

Tononi, G. (2004). An information integration theory of consciousness. *BMC Neuroscience, 5*(1), 42.

Tononi, G. (2005). Consciousness, information integration, and the brain. *Progress in Brain Research, 150*, 109–126.

Tononi, G. (2008). Consciousness as integrated information: A provisional manifesto. *The Biological Bulletin, 215*(3), 216–242.

Tononi, G. (2010). Information integration: Its relevance to brain function and consciousness. *Archives Italiennes de Biologie, 148*(3), 299–322.

Tononi, G. (2012a). The integrated information theory of consciousness: An updated account. *Archives Italiennes de Biologie, 150*(2/3), 56–90.

Tononi, G. (2012b). *Phi: A voyage from the brain to the soul*. Pantheon Books.

Tononi, G., Boly, M., Grasso, M., Hendren, J., Juel, B. E., Mayner, W. G., Marshall, W., & Koch, C. (2022). IIT, half masked and half disfigured. *Behavioral and Brain Sciences, 45*, e60.

Tononi, G., Boly, M., Massimini, M., & Koch, C. (2016). Integrated information theory: From consciousness to its physical substrate. *Nature Reviews Neuroscience, 17*(7), 450–461.

Tononi, G., & Edelman, G. M. (1998). Consciousness and complexity. *Science, 282*(5395), 1846–1851.

Tononi, G., & Koch, C. (2015). Consciousness: Here, there and everywhere? *Philosophical Transactions of the Royal Society B: Biological Sciences, 370*(1668), 20140167.

Travers, E., Frith, C. D., & Shea, N. (2018). Learning rapidly about the relevance of visual cues requires conscious awareness. *Quarterly Journal of Experimental Psychology, 71*(8), 1698–1713.

Trestman, M. (2013). The Cambrian explosion and the origins of embodied cognition. *Biological Theory, 8*(1), 80–92.

Trestman, M. (2017). Minds and bodies in animal evolution. In K. Andrews & J. Beck (Eds.), *The Routledge handbook of animals minds* (pp. 206–215). Routledge.

Trewavas, A., Baluška, F., Mancuso, S., & Calvo, P. (2020). Consciousness facilitates plant behavior. *Trends in Plant Science, 25*(3), 216–217.

Tye, M. (1995). *Ten problems of consciousness: A representational theory of the phenomenal mind.* MIT Press.

Tye, M. (2016). *Tense bees and shell-shocked crabs: Are animals conscious?* Oxford University Press.

Tye, M. (2021). Qualia. In E. N. Zalta (Ed.), *The Stanford encyclopedia of philosophy* (Fall 2021 ed.). Retrieved January 2, 2022, https://plato.stanford.edu/archives/fall2021/entries/qualia/

Vallortigara, G. (2000). Comparative neuropsychology of the dual brain: A stroll through animals' left and right perceptual worlds. *Brain and Language, 73*(2), 189–219.

Vallortigara, G., & Rogers, L. J. (2020). A function for the bicameral mind. *Cortex, 124*, 274–285.

Veit, W. (2019). Evolution of multicellularity: Cheating done right. *Biology & Philosophy, 34*(3), 34.

Veit, W. (2021a). Agential thinking. *Synthese, 199*(5), 13393–13419.

Veit, W. (2021b). The evolution of knowledge during the Cambrian explosion. *Behavioral and Brain Sciences, 44*, e47.

Veit, W. (2021c). Experimental philosophy of medicine and the concepts of health and disease. *Theoretical Medicine and Bioethics, 42*(3), 169–186.

Veit, W. (2021d). Model diversity and the embarrassment of riches. *Journal of Economic Methodology, 28*(3), 291–303.

Veit, W. (2021e). Samir Okasha's philosophy. *Lato Sensu: revue de la Société de philosophie des sciences, 8*(3), 1–8.

Veit, W. (2022a). Complexity and the evolution of consciousness. *Biological Theory.* https://doi.org/10.1007/s13752-022-00407-z

Veit, W. (2022b). Consciousness, complexity, and evolution. *Behavioral and Brain Sciences, 45*, e61.

Veit, W. (2022c). The origins of consciousness or the war of the five dimensions. *Biological Theory.* https://doi.org/10.1007/s13752-022-00408-y

Veit, W. (2022d). Scaffolding natural selection. *Biological Theory, 17*(2), 163–180.

Veit, W. (2022e). Towards a comparative study of animal consciousness. *Biological Theory.* https://doi.org/10.1007/s13752-022-00409-x

Veit, W. (2023a). *Health, agency, and the evolution of consciousness* [Ph.D. thesis]. University of Sydney.

Veit, W. (2023b). Health, consciousness, and the evolution of subjects. *Synthese.* 201, 3.

Veit, W., & Browning, H. (2020a). Perspectival pluralism for animal welfare. *European Journal for Philosophy of Science, 11*(1), 1–14.

Veit, W., & Browning, H. (2020b). Two kinds of conceptual engineering. *Preprint.* http://philsci-archive.pitt.edu/17452/

Veit, W., & Browning, H. (2021). Extending animal welfare science to include wild animals. *Animal Sentience,* 7(20), 1–4.

Veit, W. & Browning, H. (2022). Pathological complexity and the evolution of sex differences. *Behavioral and Brain Sciences, 45*, E149.

Veit, W., & Huebner, B. (2020). Drawing the boundaries of animal sentience. *Animal Sentience, 29*(13).

Veit, W., & Ney, M. (2021). Metaphors in arts and science. *European Journal for Philosophy of Science, 11*(2), 1–24.

Velmans, M. (1991). Is human information processing conscious? *Behavioral and Brain Sciences, 14*(4), 651–669.

Vonk, J. (2019). A fish eye view of the mirror test. *Learning & Behavior, 48*, 193–194.

Waggoner, B. (2003). The Ediacaran biotas in space and time. *Integrative and Comparative Biology, 43*(1), 104–113.
Walsh, D. M. (2015). *Organisms, agency, and evolution*. Cambridge University Press.
Walters, E. T. (2018). Nociceptive biology of molluscs and arthropods: Evolutionary clues about functions and mechanisms potentially related to pain. *Frontiers in Physiology, 9*, 1049.
Walters, E. T., Carew, T. J., & Kandel, E. R. (1981). Associative learning in Aplysia: Evidence for conditioned fear in an invertebrate. *Science, 211*(4481), 504–506.
Ward, J. (2013). Synesthesia. *Annual Review of Psychology, 64*, 49–75.
Watkins, N. W. (2018). (A)phantasia and severely deficient autobiographical memory: Scientific and personal perspectives. *Cortex, 105*, 41–52.
Watson, J. B. (1913). Psychology as the behaviorist views it. *Psychological Review, 20*(2), 158.
Watson, M. R., Akins, K. A., Spiker, C., Crawford, L., & Enns, J. T. (2014). Synesthesia and learning: A critical review and novel theory. *Frontiers in Human Neuroscience, 8*, 98.
Wellman, H. M. (2014). *Making minds: How theory of mind develops*. Oxford University Press.
Wells, M. (1978). *Octopus: Physiology and behaviour of an advanced invertebrate*. Chapman and Hall.
Whiten, A. (2013). Humans are not alone in computing how others see the world. *Animal Behaviour, 86*(2), 213–221.
Wigglesworth, V. B. (1980). Do insects feel pain. *Antenna, 4*, 8–9.
Wilkes, K. V. (1984). Is consciousness important? *British Journal for the Philosophy of Science, 35*, 223–243.
Woodruff, M. L. (2017). Consciousness in teleosts: There is something it feels like to be a fish. *Animal Sentience, 13*(1).
Wray, G. A. (2015). Molecular clocks and the early evolution of metazoan nervous systems. *Philosophical Transactions of the Royal Society B: Biological Sciences, 370*(1684), 20150046.
Zeman, A., Dewar, M., & Sala, S. D. (2015). Lives without imagery – Congenital aphantasia. *Cortex, 73*, 378–380.
Zuk, M. (2016). Mates with benefits: When and how sexual cannibalism is adaptive. *Current Biology, 26*(23), R1230–R1232.

INDEX

Note: *Italicized* and **bold** page numbers refer to figures and tables. Page numbers followed by "n" refer to notes.

For Product Safety Concerns and Information please contact our EU representative GPSR@taylorandfrancis.com
Taylor & Francis Verlag GmbH, Kaufingerstraße 24, 80331 München, Germany

www.ingramcontent.com/pod-product-compliance
Ingram Content Group UK Ltd.
Pitfield, Milton Keynes, MK11 3LW, UK
UKHW052225190926
475702UK00007B/1207

* 9 7 8 1 0 3 2 3 4 3 6 0 0 *